FRONTIERS IN ELECTRONICS

Proceedings of the Workshop on Frontiers in Electronics 2009

SELECTED TOPICS IN ELECTRONICS AND SYSTEMS

Editor-in-Chief: **M. S. Shur**

---

*Published**

Vol. 40: SiC Materials and Devices — Vol. 1
eds. *M. S. Shur, S. Rumyantsev and M. Levinshtein*

Vol. 41: Frontiers in Electronics
Proceedings of the WOFE-04
eds. *H Iwai, Y. Nishi, M. S. Shur and H. Wong*

Vol. 42: Transformational Science and Technology for the Current and Future Force
eds. *J. A. Parmentola, A. M. Rajendran, W. Bryzik, B. J. Walker, J. W. McCauley, J. Reifman, and N. M. Nasrabadi*

Vol. 43: SiC Materials and Devices — Vol. 2
eds. *M. S. Shur, S. Rumyantsev and M. Levinshtein*

Vol. 44: Nanotubes and Nanowires
ed. *Peter J. Burke*

Vol. 45: Proceedings of the 2006 IEEE Lester Eastman Conference on Advanced Semiconductor Devices
eds. *Michael S. Shur, P. Maki and J. Kolodzey*

Vol. 46: Terahertz Science and Technology for Military and Security Applications
eds. *Dwight L. Woolard, James O. Jensen, R. Jennifer Hwu and Michael S. Shur*

Vol. 47: Physics and Modeling of Tera- and Nano-Devices
eds. *M. Ryzhii and V. Ryzhii*

Vol. 48: Spectral Sensing Research for Water Monitoring Applications and Frontier Science and Technology for Chemical, Biological and Radiological Defense
eds. *D. Woolard and J. Jensen*

Vol. 49: Spectral Sensing Research for Surface and Air Monitoring in Chemical, Biological and Radiological Defense and Security Applications
eds. *J.-M. Theriault and J. Jensen*

Vol. 50: Frontiers in Electronics
eds. *Sorin Cristoloveanu and Michael S. Shur*

Vol. 51: Advanced High Speed Devices
eds. *Michael S. Shur and P. Maki*

Vol. 52: Frontiers in Electronics
Proceedings of the Workshop on Frontiers in Electronics 2009
eds. *Sorin Cristoloveanu and Michael S. Shur*

*The complete list of the published volumes in the series can be found at
http://www.worldscibooks.com/series/stes_series.shtml

Selected Topics in Electronics and Systems – Vol. 52

FRONTIERS IN ELECTRONICS

Proceedings of the Workshop on Frontiers in Electronics 2009

Rincon, Puerto-Rico 13 – 16 December 2009

Editors

Sorin Cristoloveanu
Grenoble INP – Minatec, France

Michael S. Shur
Rensselaer Polytechnic Institute, USA

World Scientific

NEW JERSEY • LONDON • SINGAPORE • BEIJING • SHANGHAI • HONG KONG • TAIPEI • CHENNAI

Published by

World Scientific Publishing Co. Pte. Ltd.
5 Toh Tuck Link, Singapore 596224
USA office: 27 Warren Street, Suite 401-402, Hackensack, NJ 07601
UK office: 57 Shelton Street, Covent Garden, London WC2H 9HE

British Library Cataloguing-in-Publication Data
A catalogue record for this book is available from the British Library.

Selected Topics in Electronics and Systems — Vol. 52
FRONTIERS IN ELECTRONICS
Proceedings of the Workshop on Frontiers in Electronics 2009

ISBN 978-981-4383-71-4

Printed in Singapore

PREFACE

The Workshop on Frontiers in Electronics – WOFE–09 – took place in Rincon, Puerto-Rico, in December of 2009. This meeting was the sixth in the series of the WOFE workshops (see Figure 1) and strongly reinforced the tradition of scientific quality and visionary research. Fifty leading experts from academia, industry, and government agencies came to Puerto Rico to report on the most recent and exciting developments in their fields. The warm and friendly atmosphere was ideal for inciting the participants to exchange frank and sometimes rather controversial views on future trends and directions of the electronics and photonics industry.

The conference was sponsored by the National Science Foundation, Rensselaer Polytechnic Institute, and Seoul National University. Two companies – Nanoscience, Inc. and Sensor Electronic Technology, Inc. sponsored Best Paper Award, Best Student paper Award and Innovation Paper Award. IEEE EDS provided Technical Sponsorship.

Mr. V. S. Chivukula of RPI received Best Student Paper Award for his paper entitled "Wide Band Gap Semiconductor LC Oscillator based Ultraviolet Sensors". Mr. Wu of the University of Puerto Rico received Nanoscience Best Student Paper Award for the paper entitled "Room Temperature Ferromagnetic Behavior in Yb-doped GaN Semiconductor". Best Poster Paper Award went to G. Belenky and colleaguess and the Nanoscience Best Student Poster Award was received by Mr. S. Ghosh. Professor Fernando A. Ponce received the SET innovation Award, while Nanoscience Best Paper Award went to Professor Enrique Calleja. Finally, Professor Kim M. Lewis received the Best Paper Award for the paper entitled 'Molecular Conductance: Exploring Functionality in Simple Circuits'.

This issue includes the best papers of WOFE-09 invited by the Editors and down selected after the peer review. The aim of this book was to collect the best papers of the conference, in an extended version, and make them known in the international arena. These papers are divided into four sections: advanced terahertz and photonics devices; silicon and germanium on insulator and advanced CMOS and MOSHFETs; nanomaterials and nanodevices; and wide band gap technology for high power and UV photonics. These key issues are in forefront of the microelectronics research and many papers in this book go well beyond the discussions of the original work giving a good overview of the field.

This book will be useful for nano-microelectronics scientists, engineers, and visionary research leaders. It is also recommended to graduate students working at the frontiers of the nanoelectronics and microscience.

On behalf of the WOFE Organizing, Program, and Steering Committees, we would like to thank all participants and especially the invited contributors to this issue for making WOFE–09 a successful conference. We also gratefully acknowledge generous support of this workshop by the National Science Foundation. Our special thanks go to the Members of Organizing, Program, and Steering Committees, and to Session Organizers for their tireless work and inspiration.

The next WOFE will take place in December 2011 somewhere in the Caribbean. We welcome suggestions for topics to be addressed, special sessions, and tutorials.

We believe that nanoelectronics and microelectronics are the cross roads. WOFE-2011 will provide a unique opportunity to discuss what has to be done to stimulate Nano-Micro-Electronics R&D during and after the on-going economic crisis.

This material is based upon work supported by the National Science Foundation under Grant No. 0939894.

Sorin Cristoloveanu and Michael Shur
Editors

CONTENTS

Preface v

Advanced Terahertz and Photonics Devices **1**

Broadband Terahertz Wave Generation, Detection and Coherent Control Using Terahertz Gas Photonics 3
J. Liu, J. Dai, X. Lu, I. C. Ho and X. C. Zhang

How do We Lose Excitation in the Green? 13
C. Wetzel, Y. Xia, W. Zhao, Y. Li, M. Zhu, S. You, L. Zhao, W. Hou, C. Stark and M. Dibiccari

Silicon Finfets as Detectors of Terahertz and Sub-Terahertz Radiation 27
W. Stillman, C. Donais, S. Rumyantsev, M. Shur, D. Veksler, C. Hobbs, C. Smith, G. Bersuker, W. Taylor and R. Jammy

Progress in Development of Room Temperature CW GaSb based Diode Lasers for 2-3.5 μm Spectral Region 43
T. Hosoda, J. Chen, G. Tsvid, D. Westerfeld, R. Liang, G. Kipshidze, L. Shterengas and G. Belenky

WDM Demultiplexing by Using Surface Plasmon Polaritons 51
D. K. Mynbaev and V. Sukharenko

Silicon and Germanium on Insulator and Advanced CMOS and MOSHFETs **63**

Connecting Electrical and Structural Dielectric Characteristics 65
G. Bersuker, D. Veksler, C. D. Young, H. Park, W. Taylor, P. Kirsch, R. Jammy, L. Morassi, A. Padovani and L. Larcher

Advanced Solutions for Mobility Enhancement in SOI MOSFETs 81
L. Pham-Nguyen, C. Fenouillet-Beranger, P. Perreau, S. Denorme, G. Ghibaudo, O. Faynot, T. Skotnicki, A. Ohata, M. Casse, I. Ionica, W. van den Daele, K-H. Park, S-J. Chang, Y-H. Bae, M. Bawedin and S. Cristoloveanu

Electron Scattering in Buried InGaAs/High-K MOS Channels 95
S. Oktyabrsky, P. Nagaiah, V. Tokranov, M. Yakimov, R. Kambhampati, S. Koveshnikov, D. Veksler, N. Goel and G. Bersuker

Low Frequency Noise and Interface Density of Traps in InGaAs MOSFETs with GdScO3 High-K Dielectric 105
S. Rumyantsev, W. Stillman, M. Shur, T. Heeg, D. G. Schlom, S. Koveshnikov, R. Kambhampati, V. Tokranov and S. Oktyabrsky

Low-Power Biomedical Signal Monitoring System for Implantable Sensor Applications 115
M. R. Haider, J. Holleman, S. Mostafa and S. K. Islam

Nanomaterials and Nanodevices **129**

III-V Compound Semiconductor Nanowires for Optoelectronic Device Applications 131
Q. Gao, H. J. Joyce, S. Paiman, J. H. Kang, H. H. Tan, Y. Kim, L. M. Smith, H. E. Jackson, J. M. Yarrison-Rice, J. Zou and C. Jagadish

Electron Heating in Quantum-Dot Structures with Collective Potential Barriers 143
L. H. Chien, A. Sergeev, N. Vagidov, V. Mitin and S. Birner

Electronic Structure of Graphene Nanoribbons Subjected to Twist and Nonuniform Strain 153
A. Dobrinsky, A. Sadrzadeh, B. I. Yakobson and J. Xu

Low-Frequency Electronic Noise in Graphene Transistors: Comparison with Carbon Nanotubes 161
G. Liu, W. Stillman, S. Rumyantsev, M. Shur and A. A. Balandin

ZnO Nanocrystalline High Performance Thin Film Transistors 171
B. Bayraktaroglu, K. Leedy and R. Neidhard

Zinc Oxide Nanoparticles for Ultraviolet Photodetection 183
S. Sawyer, L. Qin and C. Shing

Carbon-Based Nanoelectromechanical Devices 195
S. Bengtsson, P. Enoksson, F. A. Ghavanini, K. Engström, P. Lundgren, E. E. B. Campbell, J. Ek-Weis, N. Olofsson and A. Eriksson

Charge Puddles and Edge Effect in a Graphene Device as Studied by a Scanning Gate Microscope 205
J. Chae, H. J. Yang, H. Baek, J. Ha, Y. Kuk, S. Y. Jung, Y. J. Song, N. B. Zhitenev, J. A. Stroscio, S. J. Woo and Y.-W. Son

Wide Band Gap Technology for High Power and UV Photonics **217**

Novel Approaches to Microwave Switching Devices using Nitride Technology 219
G. Simin, J. Wang, B. Khan, J. Yang, A. Sattu, R. Gaska and M. Shur

Author Index 229

ADVANCED TERAHERTZ AND PHOTONICS DEVICES

BROADBAND TERAHERTZ WAVE GENERATION, DETECTION AND COHERENT CONTROL USING TERAHERTZ GAS PHOTONICS

JINGLE LIU, JIANMING DAI, XIAOFEI LU, I-CHEN HO and X.-C. ZHANG

Center for Terahertz Research, Rensselaer Polytechnic Institute, 110 8th Street, Troy, New York 12180, US
zhangxc@rpi.edu

Terahertz (THz) gas photonics uses gas as THz emitter and sensor for time-domain spectroscopy. Unique properties of the gas promise scalable, strong THz wave generation with broad spectral range covering the entire THz gas (0.3 THz to 35 THz). The systematic study of THz wave generation and detection in different gases shows that the generation efficiency is monotonically decreasing with the ionization potential of the gas molecules while the detection efficiency is linearly proportional to the third order nonlinear coefficient of the gas molecules. We also discuss the development of THz wave detection using laser-induced fluorescence and coherent control with THz gas photonics.

Keywords: Terahertz wave; spectroscopy; laser-induced plasma; fluorescence.

1. Introduction

With the fast development of laser technologies in the past decades, ultrafast time-domain THz spectroscopic sensing and imaging have been widely available and used in areas of semiconductor characterization, security screening, industrial inspection and communications [1-3]. Among the various THz generation and detection methods, THz gas photonics has attracted a great amount of research interest and efforts in recent years due to its broad spectral coverage and scalability [4-9]. Unlike solids, gases do not have a damage threshold, or experience phonon absorption and interface reflection. This uniqueness makes gas an ideal emitter and sensor with continuous and broadband spectral response. The only limitation for the previously reported 10 THz bandwidth is the laser pulse duration [9]. Through using even shorter and stronger laser pulses, the THz spectral range using gas photonics can be further extended to 35 THz (1% of the spectrum maximum) and the peak field can reach up to 100 kV/cm. These advances will be driving many new discoveries in fundamental molecular, biological and materials dynamics. In this paper, we will cover the state-of-art technologies in THz gas photonics, including THz air generation, detection and coherent manipulation.

2. The Mechanism of THz Generation and Detection in Gas

THz wave generation in gases has emerged as one of most promising broadband and high-intensity THz techniques. There are several ways to generate THz waves in gases such as single-color excitation, dual-color excitation and DC-biased excitation [10]. Here we focus our discussion on the THz generation scheme using dual-color laser pulse excitation which is most commonly used due to its high efficiency and controllability. In the process of THz wave generation in ambient air, one fundamental laser pulse (ω) and its second harmonic pulse (2ω) are collinearly focused into air. The second harmonic pulse is typically generated via frequency doubling by passing the fundamental pulse through a nonlinear beta barium borate crystal (BBO). The typical total optical intensity at the focus is roughly $10^{13} \sim 10^{14}$W/cm^2 or even higher. At the optical focus, the strong laser field will release electrons from the air molecules through multi-photon or tunnel ionization. As a result, intense THz pulses are emitted from the ionized air. Four-wave-mixing theory [4] and plasma transient current model [8] have been proposed to explain the physical process of THz air generation.

The reciprocal, nonlinear optical process for THz wave generation can be used to detect THz wave in gases. Dai *et al* demonstrated that a THz pulse can be detected coherently in air by using a relatively strong probe pulse [7]. Since air has to be broken down for coherent detection, this method is called THz air-breakdown coherent detection (THz-ABCD). Later Karpowicz *et al* introduced a new coherent THz detection method using an external AC bias on the probe plasma as a local oscillator [9]. This heterodyne method is called THz air-biased-coherent-detection (also abbreviated as THz-ABCD).

By referencing the modulation frequency of the AC bias to the lock-in amplifier, the measured second harmonic intensity is given by

$$I_{2\omega} \propto \left[\chi^{(3)} I_{\omega}\right]^2 E_{bias} E_{THz} \tag{1}$$

Where the $\chi^{(3)}$ is the third-order nonlinear optical coefficient of the gas and E_{bias} is the applied AC bias. Since bias-induced second harmonic intensity can be used as a local oscillator, the coherent THz detection can be realized without using the high probe energy. The typically used pulse energy is ~100 μJ, depending on the probe pulse duration and gas ionization potential. This allows one to utilize most of the laser power for the generation part to achieve a high THz field.

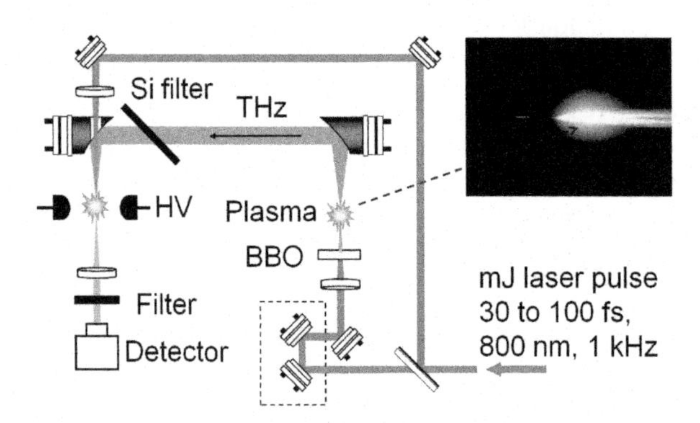

Fig. 1. The schematics of the experimental setup of broadband THz generation and detection using gas. HV, high voltage. BBO, beta barium borate. Filter, 400 nm band-pass filter. Detector, photo-multiplier tube.

Fig. 1 illustrates the typical experimental set-up of the THz air generation and air detection. Laser pulse with energy of sub mJ to a few mJ is used for THz air generation. The probe pulse is focused at the THz beam focal point. A 20 kV/cm high voltage AC bias is applied on the plasma formed by the probe pulse. The second harmonic photons measured by the photo-multiplier tube, are composed of the second harmonic photons generated through FWM of fundamental beam and THz beam, and that generated through FWM of fundamental beam and the bias field.

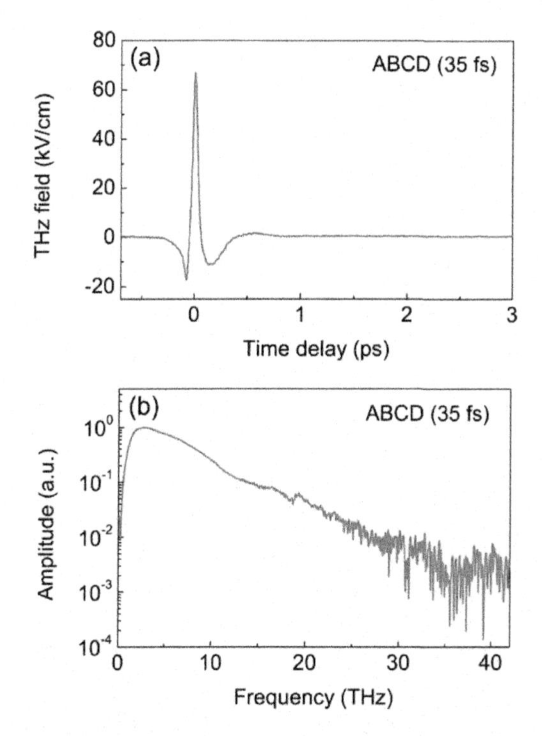

Fig. 2. (a) Typical time-domain waveform of a THz pulse using air generation and air detection. A 35 fs short laser pulse is used for generation and detection. (b) The corresponding THz spectrum in the frequency domain.

Fig. 2(a) shows a typical waveform of the THz pulse using air generation and air detection with bias. A laser pulse with 35 fs pulse duration, 600 μJ pulse energy and 800 nm center wavelength is used here. The peak THz field is about 70 kV/cm. The corresponding THz spectrum is plotted in Fig. 2(b). Due to the shorter pulse duration, the THz spectrum covers up to 35 THz (1% of the spectrum maximum). The small dip around 18.5 THz in the spectrum is due to the THz absorption in the silicon filter.

3. THz Wave Generation and Detection in Different Gases

In addition to the study of ambient air and dry nitrogen, the THz wave generation and detection has been investigated in other gases such as alkane gases and noble gases to reveal the role of the gas species in the efficiency of THz wave generation.

The experiment of THz generation in different gases was performed in a supersonic gas jet where the background pressure was kept below 10^{-5} torr. The back pressure of the gas jet is 50 torr for all the gases. The reason that a gas jet is used rather than a gas cell, is that the intensity clamping [11], self-focusing or other nonlinear optical effects during the laser propagation in gas media can be significantly reduced so that the measured THz wave generation efficiency only depends on the gas species. Electro-optical (EO) sampling is used for the measurement of the time-domain THz waveform [12]. The dependence of the peak THz field on the ionization potential of gases was measured and the results are plotted in Fig. 3.

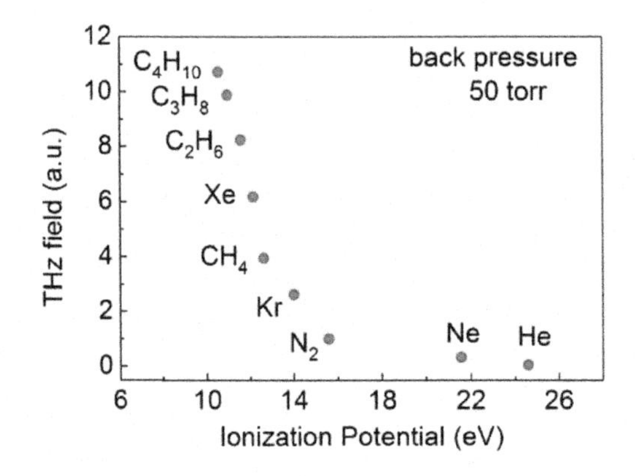

Fig. 3. Peak field amplitude of the THz pulse generated from different gases versus the ionization potential.

The results indicate that the THz wave generation efficiency monotonically decreases as the ionization potential of the gas increases. At the same gas pressure and laser intensity, the THz generation efficiency in n-butane (C_4H_{10}) whose ionization potential is 10.53 eV, is two hundred times larger than that in He whose ionization potential is 24.59 eV. The higher THz generation efficiency in the gas of lower ionization potential is attributed to the higher nonlinear coefficient $\chi^{(3)}$ in the four-wave-mixing model or larger plasma density/current in the plasma transient current model.

Broadband THz wave detection has been systematically investigated using selected gases. The dependencies of the detected second harmonic intensity on the third order nonlinear susceptibility are studied with nitrogen and alkane gases. The results are shown as blue dots in Fig. 4. The gas pressure is 100 torr for all the gases. For the sample in liquid phase, a gas dilute system is used to obtain pure vapor. The probe pulse energy is 15 μJ. A low bias field of 5 kV/cm is used to eliminate the electrical breakdown. The blue dashed line is a quadratic fitting of the data. The agreement between the measurement and the fitting further confirms Eq. (1). It has been found that C_6H_{14} generates a second harmonic signal nearly two orders higher than nitrogen.

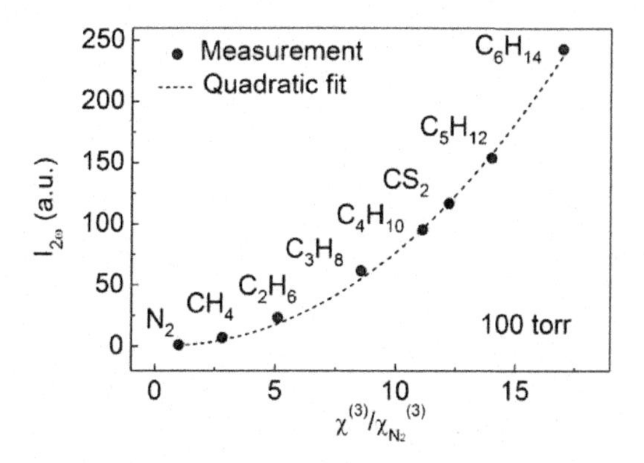

Fig. 4. The measured THz-induced second harmonic generation from laser plasma as a function of the third-order nonlinear coefficient $\chi^{(3)}$. [13]

4. THz Remote Generation and Detection

THz spectroscopy at remote distance has attracted a great amount of interest and effort due to its promising applications in areas of homeland security, environmental science and aerospace technologies. However, THz remote sensing has been a big challenge because of strong, ambient moisture absorption in the THz frequency range. Conventional THz generation and detection techniques cannot be applied to remote THz spectroscopy without suffering substantial attenuation during the THz wave propagation in the air. THz wave gas photonics provides a unique opportunity to overcome this long-existing barrier.

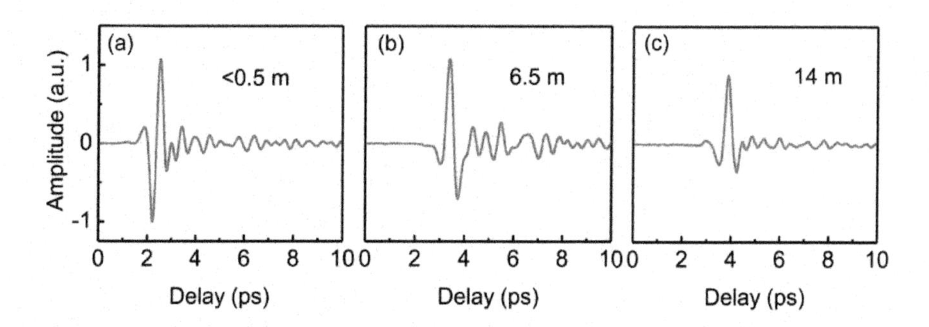

Fig. 5. THz waveforms measured at different distances. THz generation with two-color excitation is used and a 1 mm thick ZnTe crystal is used to measure the THz waveform.

In order to minimize the THz wave propagation loss in the air, plasma is created as the THz emitter by focusing two-color pulses at standoff distances [14]. The geometry of the focusing optics, the optical chirp and relative delay between two pulses are parameters which can be tuned for the optimization of THz generation.

Fig. 5 shows the time-domain THz waveforms measured by EO sampling at distances of < 0.5 m, 6.5 m and 14 m, respectively. It can be seen that by precisely controlling the focusing of the two-color pulses and their relative phase, THz pulse generation can be realized at standoff distance. In this demonstration, total pulse energy of less than 1 mJ is used for THz generation. With use of the higher-power laser, the maximal distance at which a THz wave can be generated is expected to be further extended to hundreds of meters.

With regards to THz remote detection, recently Liu *et al* introduced an omni-directional coherent THz detection technique using radiation-enhanced-emission-of-fluorescence (REEF) from gas plasma [15]. It was experimentally found and theoretically explained that the interaction between the laser-induced plasma and THz pulses leads to the enhancement of the fluorescence emission. The enhanced fluorescence depends on the THz waveform, THz field amplitude and time delay between THz pulse and plasma-generating laser pulse.

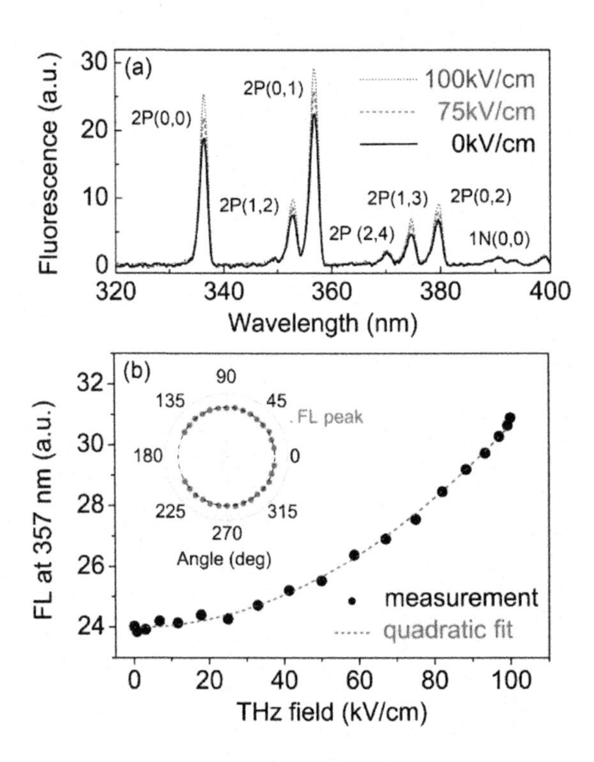

Fig. 6. (a) The nitrogen fluorescence spectra in THz fields of different strength. (b) The dependence of the fluorescence line emission at 357 nm and quadratic fit. Inset, the angular distribution of the fluorescence emission.

When the laser beam is focused into the air, the strong laser field excites or ionizes the nitrogen and oxygen molecules. The strongest fluorescence band of the nitrogen molecule lies in the spectral range of 300 nm and 400 nm. The nitrogen fluorescence spectra are measured under the influence of the THz field strengths of 0, 75 and 100 kV/cm, respectively and the results are shown in Fig. 6(a). All line emission from the second positive band (2P) and first negative band (1N) of nitrogen molecules are increased homogeneously by the THz field. The measured field dependence of the line emission at 357 nm and quadratic fitting are plotted in Fig. 6(b). The inset shows the isotropic angular distribution of the fluorescence emission. This omni-directional emission pattern allows one to measure THz waves in the backward direction, which is crucial to the remote THz wave sensing.

The mechanism for THz-REEF can be explained by the THz-wave-induced electron heating, electron-impact excitation/ionization of the high-lying Rydberg states and the subsequent increase of the fluorescing upper states. Specifically the electrons gain more kinetic energy from the THz field and then transfer part of the energy to the neighboring air molecules by collision. The more energy the electrons lose to molecules, the more upper states that can be excited. The linearity of this energy transfer process agrees well with measured quadratic dependence of the enhanced fluorescence on the THz field applied on the plasma.

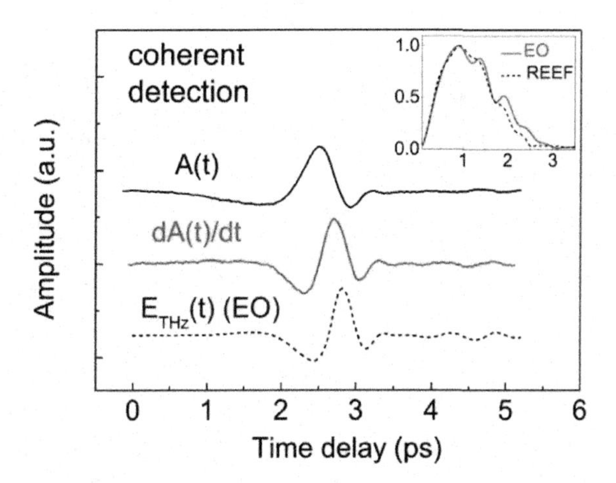

Fig. 7. The vector potential, THz waveform measured by REEF and THz waveform measured by EO, respectively. Inset shows the corresponding THz spectra by REEF and EO.

Liu *et al* also demonstrated coherent THz detection by introducing a local oscillator to the THz-plasma interaction. A 20 kV/cm AC bias with the same direction as the THz field is applied on the plasma. The AC bias is set to 500 Hz, half of the laser repetition rate, similar to THz-ABCD. The vector potential of the THz pulse can be directly measured by lock-in amplifier which is referenced to AC bias frequency. The THz field can be obtained by performing the derivative of the vector potential. Fig. 7 shows the

THz waveform measured by REEF, as described above, compared to that measured by the EO sampling. The inset shows the corresponding spectra of the two methods. The agreement between the results from the two methods indicates that THz-REEF is able to perform THz spectroscopic measurements. Further efforts could be done toward replacing the bias with the optical bias to achieve remote sensing without putting anything close to the target. By combining the THz air generation and THz-REEF, a broadband THz spectrometer with remote capability can be realized. This optical, remote THz spectroscopy is a promising and powerful means of imaging in the areas of the homeland security, environmental monitoring, etc. In addition to the application in remote sensing, THz-REEF can also be utilized in other laser plasma applications such as plasma diagnostics [16].

5. Coherent Control of THz Generation

The trajectory of the electrons ionized from the gas atoms or molecules by circularly or elliptically polarized femtosecond ω and 2ω pulses has been systematically investigated independently by two groups [17, 18]. Both groups found similar results that THz polarization rotates as the relative phase between two pulses is tuned. This provides a new approach of coherently controlling the THz wave polarization which could enable applications in fast THz wave modulation, polarization-dependent THz imaging, and coherent control of nonlinear THz interaction with materials.

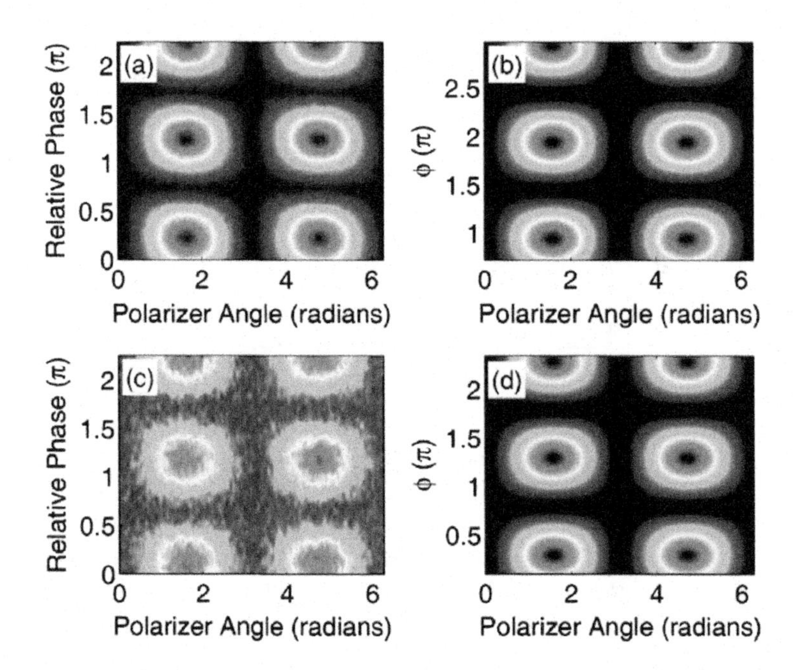

Fig. 8. (a) THz intensity versus THz polarizer angle and the relative phase between the ω and 2ω pulses with linear polarization. (a) and (c) are experimental results with the two pulses parallel and orthogonally polarized, respectively; (b) and (d) are the corresponding simulation results.

The relative phase between the ω and 2ω pulses can be continuously tuned by using an in-line phase compensator or changing the distance between the beta barium borate crystal and laser beam focus. The intensity of the THz pulse emitted from the two-color induced plasma is measured by a pyroelectric detector. Fig. 8(a) and (c) show the change in the measured THz intensity versus THz polarizer angle and relative phase between two pulses with the polarization of the ω and 2ω pulses parallel and orthogonal, respectively. Fig. 8(b) and (d) are the simulation results corresponding to the experiment in (a) and (c). A full quantum model is developed for the simulation. When the polarizations of two pulses are parallel, the second harmonic can lead to constructive interference on one side and destructive interference on the other side, which results in strong asymmetric electron velocity distribution. In the case of orthogonal polarized pulses, the weak second harmonic field only drives the electrons out of the oscillating direction of the fundamental beam, which leads to smaller, asymmetric electron velocity distribution compared to the parallel case.

6. Conclusion

Some recent progress on THz generation, detection and coherent control in gas has been briefly covered. Due to its scalability and absence of phonon absorption or interface reflection, THz wave gas photonics has been demonstrated as a powerful tool in ultra-broadband THz spectroscopy, and as an intense and polarization-controllable THz source. With further efforts in understanding the fundamental physics and then bridging it to the latest engineering technologies, THz gas photonics should find wide application in remote THz spectroscopy, nonlinear THz spectroscopy, and in-situ diagnostics with high temporal and spatial resolution.

Acknowledgement

The authors gratefully acknowledge support from the National Science Foundation, Office of Naval Research, Defense Threat Reduction Agency and the Department of Homeland Security through the DHS-ALERT Center under Award No. 2008-ST-061-ED0001. The views and conclusions contained in this document are those of the authors and should not be interpreted as necessarily representing the official policies, either expressed or implied, of the U.S. Department of Homeland Security.

References

1. M. Tonouchi, Nat. Photonics, 1, 97-105 (2007).
2. B. Ferguson and X.-C. Zhang, Nat. Mater. 1, 26-33 (2002).
3. D. Mittleman, Sensing with Terahertz Radiation (Springer, Berlin, 2003).
4. D.J. Cook and R.M. Hochstrasser, *Opt. Lett.* **25**, 1210 (2000).
5. M. Kress, T. Löffler, S. Eden, M. Thomson, and H.G. Roskos, *Opt. Lett.* **29**, 1120 (2004).
6. X. Xie, J. Dai, and X.-C. Zhang, *Phys. Rev. Lett.* **96**, 075005 (2006).
7. J. Dai, X. Xie, and X.-C. Zhang, *Phys. Rev. Lett.* **97**, 103903 (2006).

8. K.Y. Kim, J.H. Glownia, A.J. Taylor, and G. Rodriguez, *Optics Express* **15**, 4577 (2007).
9. N. Karpowicz, J. Dai, X. Lu, Y. Chen, M. Yamaguchi, H. Zhao, X.-C. Zhang, L.L. Zhang, C.L. Zhang, M. Price-Gallagher, C. Fletcher, O. Mamer, A. Lesimple, and K. Johnson, *Appl. Phys. Lett.* **92**, 011131 (2008).
10. M.D. Thomson, M. Kreβ, T. Loffer, and H.G. Roskos, *Laser & Photon. Rev.* **1**, 349 (2007).
11. A. Becker, N. Akozbek, K. Vijayalakshmi, E. Oral, C.M. Bowden, and S.L. Chin, *Appl. Phys. B* **73**, 287 (2001).
12. Q. Wu, M. Litz, and X.-C. Zhang, *Appl. Phys. Lett.* **68**, 2924 (1996).
13. D.P. Shelton, *Phys. Rev. A* **42**, 2578 (1990).
14. J. Dai, J. Liu, and X.-C. Zhang, *IEEE J. Sel. Topics Quantum Electron.* (invited) (2010).
15. J. Liu and X.-C. Zhang, *Phys. Rev. Lett.* **103**, 235002 (2009).
16. J. Liu and X.-C. Zhang, *Appl. Phys. Lett.* **96**, 041505 (2010).
17. J. Dai, N. Karpowicz, and X.-C. Zhang, *Phys. Rev. Lett.* **103**, 023001 (2009).
18. H. Wen and A.M. Lindenberg, *Phys. Rev. Lett.* **103**, 023902 (2009).

HOW DO WE LOSE EXCITATION IN THE GREEN?

CHRISTIAN WETZEL*, YONG XIA, WEI ZHAO, YUFENG LI, MINGWEI ZHU, SHI YOU, LIANG ZHAO, WENTING HOU, CHRISTOPH STARK, MICHAEL DIBICCARI

Future Chips Constellation and Department of Physics, Applied Physics and Astronomy, Rensselaer Polytechnic Institute, Troy, New York, U.S.A.

KAI LIU, MICHAEL S. SHUR

Department of Physics, Applied Physics and Astronomy, Rensselaer Polytechnic Institute, Troy, New York, U.S.A.

GREGORY A. GARRETT, MICHAEL WRABACK

Sensors and Electron Devices Directorate, U.S. Army Research Laboratory, Adelphi, Maryland 20783, U.S.A.

THEERADETCH DETCHPROHM

Future Chips Constellation and Department of Physics, Applied Physics and Astronomy, Rensselaer Polytechnic Institute, Troy, New York, U.S.A.

Efficiency droop and green gap are terms that summarize performance limitations in GaInN/GaN high brightness light emitting diodes (LEDs). Here we summarize progress in the development of green LEDs and report on time resolved luminescence data of polar c-plane and non-polar m-plane material. We find that by rigorous reduction of structural defects in homoepitaxy on bulk GaN and V-defect suppression, higher efficiency at longer wavelengths becomes possible. We observe that the presence of donor acceptor pair recombination within the active region correlates with lower device performance. To evaluate the aspects of piezoelectric polarization we compare LED structures grown along polar and non-polar crystallographic axes. In contrast to the polar material we find single exponential luminescence decay and emission wavelengths that remain stable irrespective of the excitation density. Those findings render high prospects for overcoming green gap and droop in non-polar homoepitaxial growth.

Keywords: light emitting diode; GaInN/GaN; green spectrum; efficiency; homoepitaxy; luminescence; time resolved spectroscopy; cathodoluminescence; transmission electron microscopy.

1. Introduction

Green gap has become a term to describe the challenge to achieve visible green light by any efficient means. Replacement of energy wasteful incandescent light bulbs with solid-state lighting in the form of light emitting diodes (LEDs) has lead to a rapid progress in light generation efficiency and total light output power, yet, this has primarily been limited to the AlGaInN blue[1] and the AlGaInP red[2] spectral regions. In between, despite

* Electronic mail: wetzel@ieee.org

steady progress, the green is still trailing its spectral neighbors. For example, in 430 nm AlGaInN blue, external wall plug efficiencies up to 60% have been achieved, while 525 nm AlGaInN green hardly reaches up to 27%.[3,4] This challenge in the green is compounded by the fact of an efficiency droop, common to all AlGaInN LEDs, where the light output efficiency drops substantially as the current density in the devices increases beyond some 10 A/cm^2. In typical 1 W-LEDs operate at 35 A/cm^2 (350 mA in 1 mm^2 dies), efficiency may be only half of its maximum value. The core of our work – summarized here – aims at identifying the mechanisms that lead to either or both limiting effects and at methods to overcome those universal challenges.

2. Material and Methods

As typical for high performance group-III nitride AlGaInN LEDs, our structures have been grown in metal organic vapor phase epitaxy (MOVPE), primarily along the c-axis of the uniaxial crystal structure on *c*-plane sapphire.[5] Epitaxial growth on this mismatched substrate is initiated by low-temperature deposited nucleation layers of either AlN or GaN. Over the course of some 4 μm GaN, n-doped by Si (n = ~2x10^{18}cm^{-3}) for the most part, defect densities are typically reduced to the 10^9 cm^{-2} in threading dislocations (Figure 1). The active region ensues next comprising 5 to 10 periods of typically 3 nm-wide $Ga_{1-x}In_xN$/GaN quantum wells (QWs) embedded in nominally undoped 10 to 24 nm wide GaN barriers. InN fractions x were determined by combining $\Theta/2\omega/2\theta$ x-ray diffraction scans of the superlattice structure with layer thickness information as obtained in high resolution transmission electron microscopy (TEM). For multiple QW samples, growth was terminated after the active region, while for full LED structures, an $Al_yGa_{1-y}N$ (y = ~0.07) electron blocking layer p-doped (p = ~ 10^{17}cm^{-3}) by Mg[6] at a thickness of ~15 nm and a p-GaN layer of ~180 nm (p = ~ 5x10^{17}cm^{-3}) were added.[7,8]

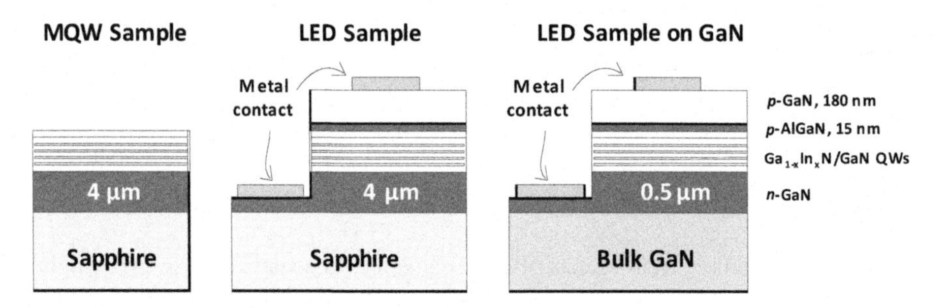

Fig. 1. Schematics of the different samples types employed.

For homoepitaxy on bulk GaN, hydride vapor phase epitaxy (HVPE) grown GaN was cut and mechano-chemically polished to epiready surfaces by collaboration partner Kyma Technologies, Raleigh, NC.[9] By virtue of the high growth rate, layers sufficiently thick to be sliced along various crystallographic planes have been obtained. This unique approach allows us to compare LED performance on the polar *c*-plane, with those on the non-polar *m*- and *a*-planes. After HVPE growth of several hundred micrometer, the threading dislocation density along the *c*-growth axis can reliably be reduced into the mid 10^6 cm^{-2}.[9]

Here we present results of materials and device performance characterization in time resolved photoluminescence (PL) spectroscopy and depth-resolved low-temperature cathodoluminescence spectroscopy.

3. Experimental

Over the course of a number of development stages in our work, avoidance of spatial GaInN alloy fluctuations and any structural defects has repeatedly proven relevant to performance improvements in particular in green GaInN/GaN QW LEDs.[10] As characterized in atomic force microscopy, scanning electron microscopy, and low temperature cathodoluminescence mapping, growth on low-dislocation density bulk GaN can boost 420 nm LED light output power tenfold over that of typical sapphire-based growth with dislocation densities in the mid 10^9 cm^{-2}.[11] Of particular concern proved to be the interaction of threading dislocations with the In-containing QWs forming V-defects that with further growth open into hexagonal pits with higher index vicinal growth surfaces. In the vicinity of such growth surface disruptions, In-incorporation and growth rate of well and barriers are found to be disturbed and, in the case of large V-defects (200 nm across), leading to very thin secondary QWs on the vicinal surfaces.[12] This is accompanied by progressively wider QWs and barriers on the remaining *c*-plane surfaces nearby. This can be understood as a result of growth rate differences of the planes involved and an effective mass transport from the defects to the *c*-plane surfaces. Hangleiter *et al.* described such formation in short wavelength blue QWs and concluded that the higher energy QWs could provide a carrier isolation from non-radiative recombination in the threading dislocation core of the V-defect.[13] For the green spectral region in particular, however, we find, that this non-uniformity leads to a rapid decay of crystalline perfection and homogeneity of the later grown QWs and, along with it, a strong decay of light output performance.[14] Not until the generation of such V-defects was effectively suppressed by proper optimization of the growth processes to promote a preferential planar growth on the *c*-plane over vicinal plane growth, the decoration of threading dislocations with V-defect could essentially be suppressed in blue, cyan, green, and even yellow LEDs.[15]

By consequential execution of the approach, light output power in respective LEDs throughout the green gap could essentially be doubled and tripled (see Figure 2).

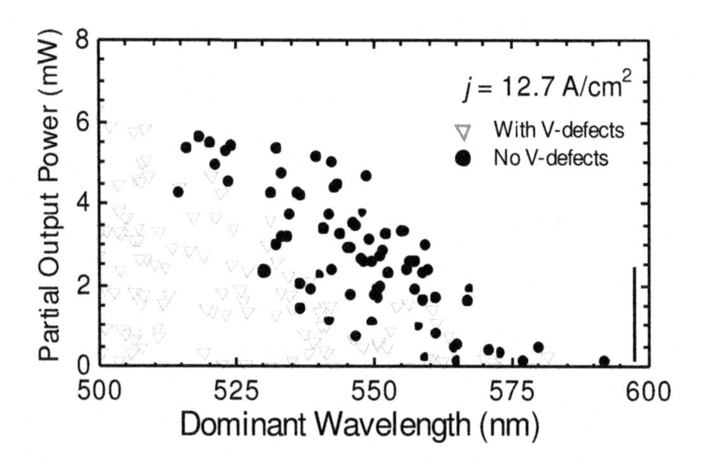

Fig. 2. Achieved light output power versus dominant wavelength of the center portion of green GaInN/GaN LED wafers of individual epi runs as measured through the substrate. Wafers without V-defect formation exhibit consistently substantial higher light output performance.

Evidence, particularly in the development of green GaInN/GaN QW LEDs has repeatedly shown that photoluminescence (PL) intensity is only a weak predictor for electroluminescence (EL) light output power performance. Nevertheless, time-resolved PL decay can well serve to assess the dominant path of recombination after high density excitation. In a sequence of MQW samples at room temperature, we observe the following correlations (Figure 3): As frequently observed in such structures, the luminescence decay cannot be explained by a single decay rate, but rather follows a stretched exponential decay behavior.[16] With increasing emission wavelength, the luminescence decay times increase rapidly from the ps range in the blue to the ns range in the green. Furthermore, the overall PL intensity directly scales with the luminescence decay time in the range of 30 ps to 1.4 ns. Both, the trend and the timescale of the decay time are well within expectation for a luminescence decay that is dominated by non-radiative recombination.[17]

The possibility to establish QW growth along both the polar and non-polar axes of GaN provides insights into the many aspects controlled by the piezoelectric polarization. In particular, since piezoelectric polarization should induce a strong electric field within the *c*-axis QW and change quantization and selection rules. This could reasonably play a crucial role in efficiency droop and green gap.

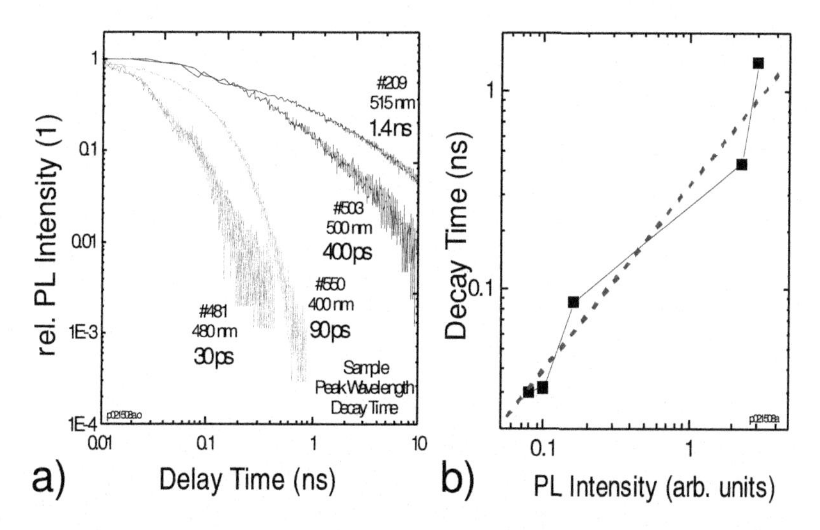

Fig. 3. a) Time resolved photoluminescence of polar *c*-plane GaInN/GaN MQW structures of different peak wavelength. b) The decay times scales well with the PL intensity indicating a dominance of non-radiative recombination.

We find that under certain growth regimes, non-polar *m*-plane and polar *c*-plane QWs can be grown simultaneously with similar growth rate. Figure 4 shows the results of two such grown QW structures on *m*-plane and *c*-plane GaN, respectively.[18] While the polar structure shows a 558 nm deep green emission in cathodoluminescence, the non-polar sister sample remains a 488 nm royal blue. A common interpretation for non-polar structures exhibiting such unexpected short emission wavelength is a reduced In-incorporation along such growth orientation.[19,20] A detailed x-ray diffraction analysis of both lattice constants in our material, however, reveals that there is no relevant difference in InN fraction in both samples that could account for such a large 320 meV discrepancy.

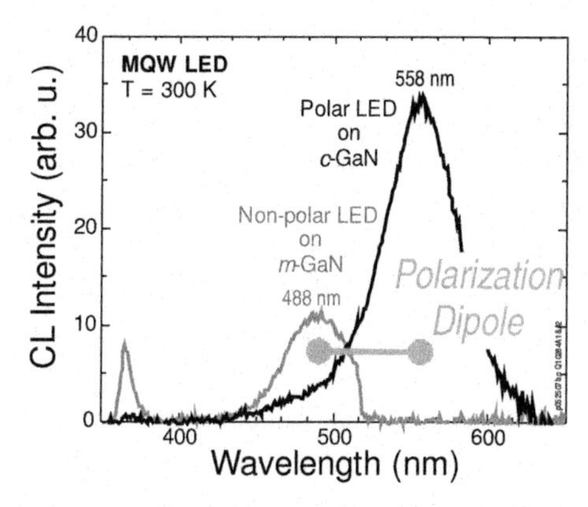

Fig. 4. Luminescence of MQW LED structures grown simultaneously along the polar *c*-axis and non-polar *m*-axis of GaN. The discrepancy in emission energy is attributed to the polarization dipole in the polar sample (after Ref. 18).

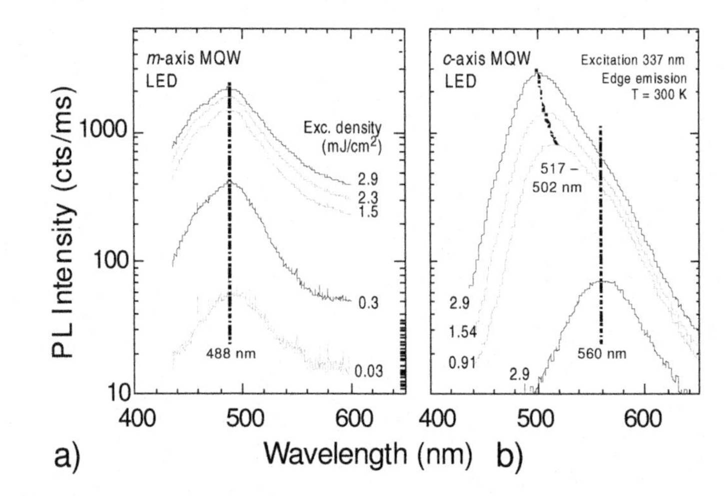

Fig. 5. PL of a) non-polar m-plane and b) polar *c*-plane GaInN/GaN MQW structure under variable high fluence optical excitation. While emission of the polar structure shows a strong blue shift, emission of the non-polar one stays put at 488 nm (after Ref. 18).

Both structures are further analyzed under variable high density pulsed 337 nm laser excitation (Figure 5). Under conditions that can lead up to stimulated in 405 nm polar *c*-plane structures, the 558 nm sample shows a transition to a dominant short wavelength emission at 502 nm.[18] This blue shift is commonly attributed to the screening of the quantum confined Stark effect in the piezoelectric *c*-plane QW. In contrast, in the non-polar structure, the emission wavelength stays pegged at 488 nm, independently of the excitation power. The advantage of longer wavelength emission in polar *c*-plane material obviously is limited to the low excitation density regime. Under high excitation, both, the polar and non-polar structures behave very similar. This behavior is in line with our earlier findings that the polarization dipole defined by the product of the net polarization charges at the QW interfaces with the QW width helps to reach the longer emission wavelengths in the polar structures. In absence of the piezoelectric polarization in the non-polar geometry, the dipole does not extend the wavelength of the QW. The magnitude of the polarization dipole can well be extrapolated from our earlier data[21,22] and should lie around 380 meV for samples of such emission wavelength. The actual observed discrepancy in the simultaneously grown samples is in rather good agreement.

Homoepitaxial growth on the various prepared surface of *c*-axis grown HVPE GaN provides the additional benefit of virtually suppressing all propagation of treading dislocations after the change of the growth direction. We have learned to grow full QW LED structures along all three of the *c*-, *a*-, and *m*-axes.[7,8,11,18] For emission wavelengths up to 485 nm in *m*-plane, 538 nm in *a*-plane but below 525 nm in *c*-plane, we can achieve this in virtual absence of threading dislocations as judged from cross-sectional transmission electron microscopy over wide sample ranges (a quantitative assessment of density $< 10^{8}$ cm^{-2} is not possible in this method). Some dislocation lines remain at a low density below 10^{5} cm^{-1}.

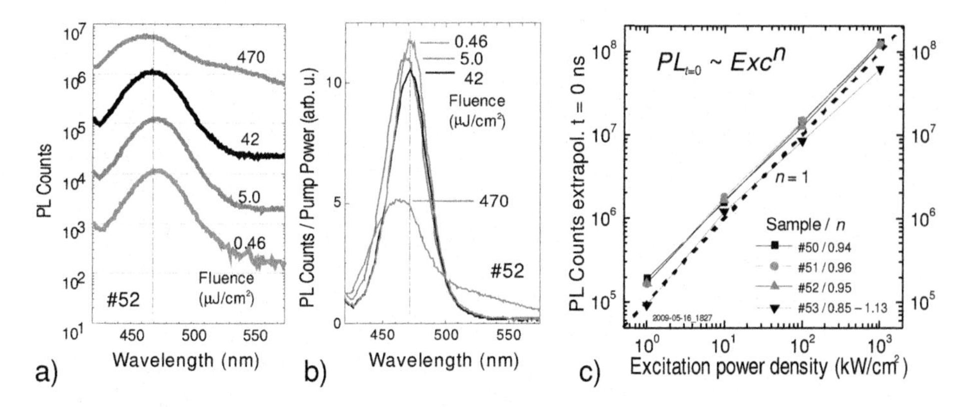

Fig. 6. PL of non-polar *m*-plane GaInN/GaN MQW structure under variable high fluence optical excitation: a) spectra show a peak wavelength that does not shift with excitation density; b) spectra normalized to the excitation fluence show little spectral variation; c) integrated PL intensity versus excitation density.

High excitation density pulsed PL in such *m*-plane MQW samples is shown in Figure 6. As the excitation fluence is increased from 0.46 up to 470 $\mu J/cm^2$, the emission spectrum undergoes only minimal variation (Figure 6a). Figure 6b) shows the spectra after normalization to the excitation fluence. The integrated luminescence output versus excitation power density (Figure c), shows a very close to linear correlation.

The time resolved PL decay of the same structures is shown in Figure 7. As the excitation fluence is increased over four orders of magnitude, the luminescence decay shows only minimal variation in the decay behavior (Figure 7a). The spectral variation of the decay times across the set of samples (Figure 7b) shows simple single exponential decay described by decay times varying only between 100 to 200 ps. Similar to the case of the polar structures, this behavior is possibly dominated by non-radiative recombination.

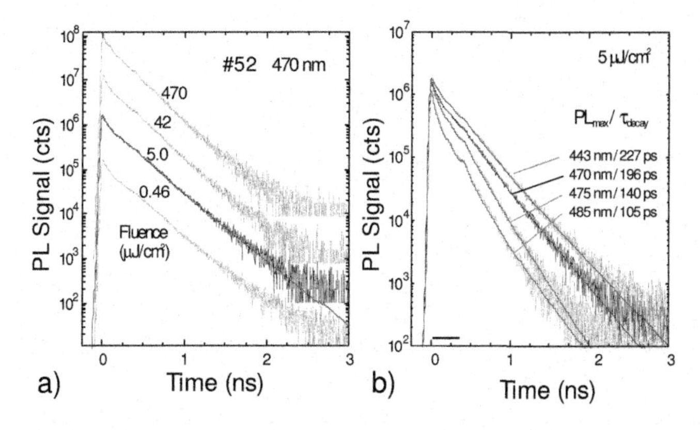

Fig. 7. Time resolved luminescence decay in non-polar *m*-plane GaInN/GaN MQW structures: a) a single exponential decay is observed independent of excitation fluence; b) as a function of center wavelength, only a minimal variation of the decay time is observed.

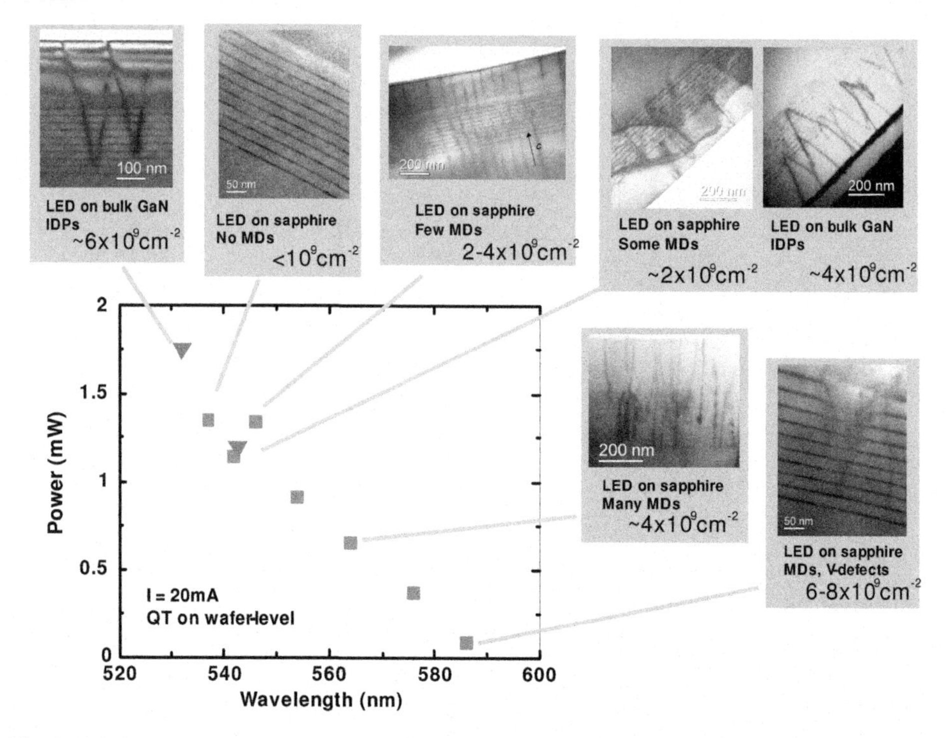

Fig. 8. Light output power in various LEDs as a function of emission wavelength along with characteristic cross-sectional TEM micrographs o defects forming in the MQW region. Listed also are the estimated dislocation densities. Triangles indicate homoepitaxial samples, squares indicate samples on sapphire substrate.

The avoidance of threading dislocations in the homoepitaxial growth of GaN on the non-polar planes of bulk GaN does not automatically exclude defects from forming in the QW active regions. In fact we find evidence, that particularly for the longer emission wavelengths, i.e. > 515 nm, new threading dislocations are being initiated in the highly strained QWs.[7,23] Figure 8 shows the light output power as a function of wavelength in a series of green and deep green LEDs. Associated with the data also are cross-sectional TEM micrographs revealing some sample-characteristic dislocation networks. The data includes samples on sapphire (red symbols) and on bulk GaN (green symbols).

It can be seen that while the presence of threading dislocations has not been suppressed in structures on sapphire, the generation of V-shaped defects is relegated to the very long wavelength range as seen in a 589 nm yellow-orange structure.

While reduction of defects observable in TEM has resulted in significant performance improvements, particularly in the green spectral region, it is doubtful, that dislocations are the only culprit for non-radiative recombination. This becomes particular apparent from the persisting droop performance in those non-polar samples, where threading dislocation densities have been pushed below the detection limit of 10^8 cm^{-2}.[7,8]

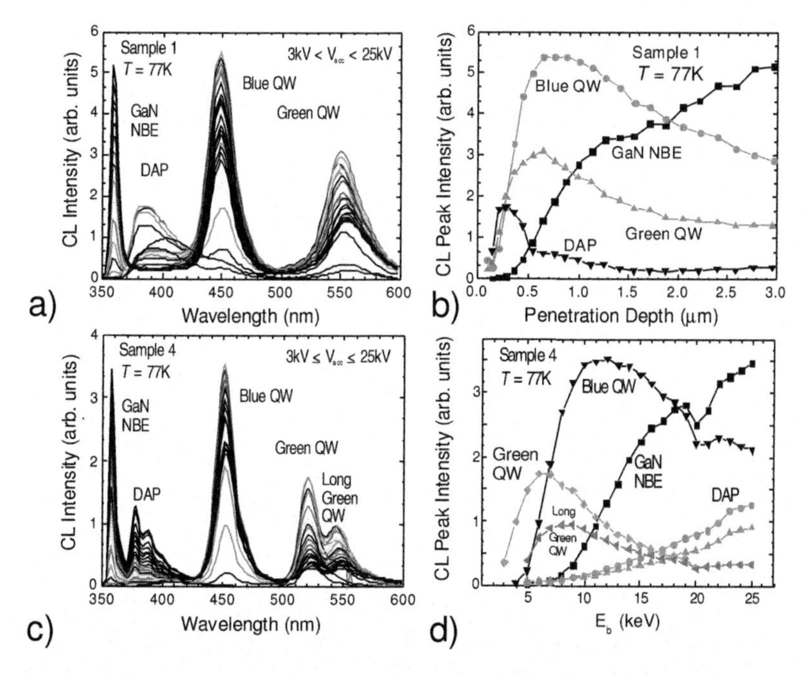

Fig. 9. Depth resolved cathodoluminescence at low temperature in a low performing (a) and b)) and a high performing (c) and d)) green LED. Spectra reveal donor- acceptor pair recombination either within the QW region (b) or deep in the n-layer (d). Graphs b) and d) are excitation depth interpretations of the acceleration voltages of spectra in a) and c) (after Ref. 24).

Cathodoluminescence of a series of green LED samples provided by Samsung Electromechanics reveals strong contributions of donor-acceptor pair recombination at low temperature. Depth resolved CL spectra of a high and a low performing sample are shown in Figure 9 a) and c).[24] Depth resolution was provided by a stepwise variation of the acceleration voltage from 3 kV for low mean deposition depth to 25 kV for high penetration. Maxima corresponding to the GaN near bandedge emission, the donor acceptor pair recombination, a blue emitting QW and one or even two different green emitting QWs are identified. In electroluminescence only the longest emission green QW is apparent independent of the sample side of light collection. This general observation underscores the important distinction of electrical from photo or electron beam excitation. The respective peak intensities as a function of excitation voltage, or its corresponding excitation depth, is shown in Figure 9 b) and d) for both samples, respectively.

The sequence of QWs can directly be identified from the sequence of respective CL peak maxima. The green QWs lie closest to the p-side and top surface layers, while the blue QWs follow at greater depth followed then by the GaN near bandedge emission associated with the n-layers of the structures.

It is important to note, that the donor acceptor pair band only appears within the n-layers in the sample with the higher light output performance, while there is a pronounced maximum in the vicinity of the green QWs in the case of the poor performing LED structure.

4. Discussion

While this correlation is not necessarily a causal one, the presence of luminescence of donor acceptor pair recombination at low temperature in a region that also carrier the desired electroluminescence is very likely to be a performance detriment. While the rate of carriers participating in luminescent donor acceptor pair recombination may be small, the channel of non-radiative recombination through the same defects is a substantially larger portion. Most importantly, it persists at room temperature, where the smaller radiative portion may be suppressed entirely. Therefore, the identification of such a loss mechanism at low temperature indeed does indicate a major channel of efficiency loss at room temperature. The argument that such recombination should be rate limited and not play a significant role under high electrical injection cannot really be upheld since room temperature defect luminescence should be the consequence, which, however is not observed.

In an attempt to identify the mechanisms of performance loss, another piece of experimental evidence needs to be accounted for. We find experimental evidence, that sub bandgap green light propagating in GaInN/GaN QW LED structures can experience a relevant optical absorption in the regime of a high light flux density.[25] An initial analysis suggests that the effect should be relevant under conditions of laser diode operation, but not necessarily limit the performance of LEDs.

Another mechanism that is being considered in the literature is non-radiative Auger recombination.[26] In this scenario, reduced transition matrix elements in QWs grown along the polar c-axes should result in a charge accumulation in the QWs. This in turn should induce higher order carrier-carrier interaction that should possibly reach the threshold for the non-radiative recombination path. The analysis of this possibility relies on a number of assumptions to assess that actual carrier concentration in the QWs and therefore is subject to a wide range of uncertainty.

A further argument revisits the model of carrier overflow, in particular, overflow of the higher mobility electrons into the p-doped layers. Early-on, such arguments lead to the introduction of the AlGaN electron blocking layer.[6,27] Experiments by the present group, however, finds that an electron blocking layer in its various modifications has only a very small impact on the performance limitations and the omnipresent droop. There is an additional possibility that the piezoelectric mismatch of the AlGaN blocking layer with the GaN barriers leads to a polarization-induced charge accumulation layer and so give rise to non-radiative Auger recombination.

Common to the majority of those alternate considerations is the role of piezoelectric polarization as a leading cause. An approach to suppress piezoelectric polarization in non-polar QW growth therefore should serve as a crucial test to verify the respective role in the green gap and efficiency droop. While the level of crystalline perfection achieved in such non-polar growth is very high, elimination of efficiency droop and green gap cannot be claimed at present. Given the very limited overall volume of LED epi growth in such non-polar crystal orientation, shortcomings likely due to inappropriate doping levels cannot be excluded. Therefore, a proof of the role of piezoelectric polarization in green gap and efficiency droop cannot be presented at this early stage. Data to access the relevant transport properties is presently not sufficiently developed to draw relevant conclusions.

5. Conclusions

The issues of green gap and efficiency droop are being addressed by the stepwise elimination of the most plausible causes, namely high structural defect density and piezoelectric polarization. Disproportionate alloy homogeneities in GaInN alloys had been eliminated already in earlier approaches of this team. In homoepitaxy on low-dislocation density bulk GaN substrates along the polar *c*- and nonpolar *a*- and *m*-axes, high quality QW LED growth has been established. In particular by virtue of eliminating piezoelectric polarization in the non-polar a- and m-axes structures, the role of piezoelectric polarization in those omnipresent performance limitations are being explored. Overall device performance in these structures, however, is not yet at a level where a discriminating call should be made. Relevant progress, however, has readily been achieved in the development of wavelength-stable green LEDs: in absence of piezoelectric polarization the emission wavelength is maintained independent of the drive current. This proves a major advantage in technical deployment of such LEDs for multi component red-green-blue solid state lighting.

Acknowledgments

This work was supported by a DOE/NETL Solid-State Lighting Contract of Directed Research under DE-EE0000627 (Brian Dotson). This work was supported by the United States Air Force AFRL/SNH under FA8718-08-C-0004 (David Bliss). This work was also supported by the National Science Foundation (NSF) Smart Lighting Engineering Research Center (# EEC-0812056).

References

1. I. Akasaki and H. Amano, Breakthroughs in Improving Crystal Quality of GaN and Invention of the p–n Junction Blue-Light-Emitting Diode, *Jpn. J. Appl. Phys.* **45**, 9001 (2006).
2. M.R. Krames, M. Ochiai-Holcomb, G.E. Höfler, C. Carter-Coman, E.I. Chen, I.-H. Tan, P. Grillot, N.F. Gardner, H.C. Chui, J.-W. Huang, S.A. Stockman, F.A. Kish, M.G. Craford, T.S. Tan, C.P. Kocot, M. Hueschen, J. Posselt, B. Loh, G. Sasser, and D. Collins, High-power truncated-inverted-pyramid $(Al_xGa_{1-x})_{0.5}In_{0.5}P/GaP$ light-emitting diodes exhibiting >50% external quantum efficiency, *Appl. Phys. Lett.* 75, 2365 (1999).
3. G. Chen, M. Craven, A. Kim, A. Munkholm, S. Watanabe, M. Camras, W. Götz, and F. Steranka, Performance of high-power III-nitride light emitting diodes, *Phys. Stat. Sol. (a)* **205**(5), 1086–1092 (2008).
4. C. Wetzel, Yufeng Li, J. Senawiratne, Mingwei Zhu, Yong Xia, S. Tomasulo, P.D. Persans, Lianghong Liu, D. Hanser, and T. Detchprohm, Characterization of GaInN/GaN layers for green emitting laser diodes, *J. Cryst. Growth* **311**, 2942–2947 (2009).

5. C. Wetzel, P. Li, T. Detchprohm, and J.S. Nelson, Optimization of Green and Deep Green GaInN/GaN Light Emitting Diodes, *Phys. Stat. Sol.* (c), **2**(7) 2871-3 (2005).
6. H Amano, M Kito, K Hiramatsu, and I Akasaki, P-type conduction in Mg-doped GaN treated with low-energy electron beam irradiation (LEEBI), *Jpn. J. Appl. Phys.* **28** L2112-L2114 (1989).
7. Theeradetch Detchprohm, Mingwei Zhu, Yufeng Li, Liang Zhao, Shi You, Christian Wetzel, Edward A. Preble, Tanya Paskova, and Drew Hanser, Wavelength-Stable Cyan and Green Light Emitting Diodes on Non-Polar m-Plane GaN Bulk Substrates, *Appl. Phys. Lett.* **96**(5), 051101 (2010).
8. Theeradetch Detchprohm, Mingwei Zhu, Yufeng Li, Yong Xia, Christian Wetzel, Edward A. Preble, Lianghong Liu, Tanya Paskova, and Drew Hanser, Green Light Emitting Diodes on a-Plane GaN Bulk Substrates, *Appl. Phys. Lett.* **92**, 24119 (2008).
9. D. Hanser, M. Tutor, E. Preble, M. Williams, X. Xu, D. Tsvetkov, and L. Liu, TILTE NEEDED *J. Cryst. Growth* **305**, 372 (2007).
10. C. Wetzel, T. Salagaj, T. Detchprohm, P. Li, and J.S. Nelson, GaInN/GaN Growth Optimization for High Power Green Light Emitting Diodes, *Appl. Phys. Lett.* **85**(6), 866-8 (2004).
11. T. Detchprohm, Y. Xia, Y. Xi, M. Zhu, W. Zhao, Y. Li, E.F. Schubert, L. Liu, D. Tsvetkov, D. Hanser, and C. Wetzel, Dislocation Analysis in Homoepitaxial GaInN/GaN Light Emitting Diode Growth, *J. Crystal Growth* **298**, 272–275 (2007).
12. X.H. Wu, C.R. Elsass, A. Abare, M. Mack, S. Keller, P.M. Petroff, S.P. DenBaars, J.S. Speck, and S.J. Rosner, Structural origin of V-defects and correlation with localized excitonic centers in InGaN/GaN multiple quantum wells, *Appl. Phys. Lett.* **72**, 692 (1998).
13. A. Hangleiter, F. Hitzel, C. Netzel, D. Fuhrmann, U. Rossow, G. Ade, and P. Hinze, Suppression of Nonradiative Recombination by V-Shaped Pits in GaInN/GaN Quantum Wells Produces a Large Increase in the Light Emission Efficiency, *Phys. Rev. Lett.* **95**, 127402 (2005).
14. M. Zhu, T. Detchprohm, S. You, Y. Wang, Y. Xia, W. Zhao, Y. Li, J. Senawiratne, Z. Zhang, and C. Wetzel, V-defect Analysis in Green and Deep Green Light Emitting Diode Structures, *Phys. Stat. Sol. (c)*, **5**(6), 1777–1779 (2008).
15. T. Detchprohm, M. Zhu, Y. Xia, Y. Li, W. Zhao, J. Senawiratne, and C. Wetzel, Improved Performance of GaInN Based Deep Green Light Emitting Diodes through V-Defect Reduction, *Phys. Stat. Sol. (c)* **5**(6), 2207–2209 (2008).
16. M. Pophristic, F.H. Long, C. Tran, I.T. Ferguson, and R.F. Karlicek, Time-resolved photoluminescence measurements of InGaN light-emitting diodes, *Appl. Phys. Lett.* **73**, 3550 (1998).
17. T. Li, A.M. Fischer, Q.Y. Wei, F.A. Ponce, T. Detchprohm, and C. Wetzel, Carrier localization and non-radiative recombination in yellow emitting InGaN quantum wells, *Appl. Phys. Lett.* **96**, 031906 (2010).
18. C. Wetzel, M. Zhu, J. Senawiratne, T. Detchprohm, P.D. Persans, L. Liu, E.A. Preble, and D. Hanser, Light Emitting Diode Development on Polar and Non-Polar GaN Substrates, *J. Cryst. Growth* **310**, 3987-91 (2008).
19. Kwang-Choon Kim, Mathew C. Schmidt, Hitoshi Sato, Feng Wu, Natalie Fellows, Makoto Saito, Kenji Fujito, James S. Speck, Shuji Nakamura, and Steven P. DenBaars, Improved electroluminescence on nonpolar m -plane InGaN/GaN quantum wells LEDs, *Phys. Stat. Sol. (RRL)* **1**(3), 125–127 (2007).
20. S.F. Chichibu, H. Yamaguchi, L. Zhao, M. Kubota, T. Onuma, K. Okamoto, and H. Ohta, Improved characteristics and issues of m-plane InGaN films grown on low defect density m-plane freestanding GaN substrates by metalorganic vapor phase epitaxy, *Appl. Phys. Lett.* **93**, 151908 (2008).

21. C. Wetzel, T. Takeuchi, H. Amano, and I. Akasaki, Electric Field Strength, Polarization Dipole, and Multi-Interface Bandoffset in GaInN/GaN Quantum Well Structures, *Phys. Rev. B* **61**, 2159-63 (2000).
22. C. Wetzel, T. Takeuchi, H. Amano, and I. Akasaki, Quantized States in $Ga_{1-x}In_xN$/GaN Heterostructures and the Model of Polarized Homogeneous Quantum Wells, *Phys. Rev. B* **62**(20), R13302-5 (2000).
23. Mingwei Zhu, Shi You, Theeradetch Detchprohm, Tanya Paskova, Edward A. Preble, Drew Hanser, and Christian Wetzel, Inclined Dislocation Pair Relaxation Mechanism in Homoepitaxial Green GaInN/GaN Light Emitting Diodes, *Phys. Rev. B* **81**, 125325 (2010).
24. Yong Xia, Yufeng Li, Theeradetch Detchprohm, and Christian Wetzel, Depth profile of donor-acceptor pair transition revealing its effect on the efficiency of green LEDs, *Physica B* **404**, 4899–4902 (2009).
25. W. Zhao, M. Zhu, Y. Xia, Y. Li, J. Senawiratne, S. You, T. Detchprohm, and C. Wetzel, Very Strong Nonlinear Optical Absorption in Green GaInN/GaN Multiple Quantum Well Structures, *Phys. Stat. Sol. (b)* **245**(5), 916–919 (2008).
26. Y.C. Shen, G.O. Müller, S. Watanabe, N.F. Gardner, A. Munkholm, and M.R. Krames, Auger recombination in InGaN measured by photoluminescence, *Appl. Phys. Lett.* **91**, 141101 (2007).
27. Isamu Akasaki, and Hiroshi Amano, Perspective of the UV/blue light emitting devices based on GaN and related compounds, *Optoelectron. Devices Technol.* **7**, (1), 49–56 (1992).

SILICON FINFETS AS DETECTORS OF TERAHERTZ AND SUB-TERAHERTZ RADIATION

W. STILLMAN

Department of Electrical, Computer and Systems Engineering,
Rensselaer Polytechnic Institute, Troy NY 12180, USA
stillw2@rpi.edu

C. DONAIS, S. RUMYANTSEV and M. SHUR

Rensselaer Polytechnic Institute, Troy NY 12180, USA

D. VEKSLER, C. HOBBS, C. SMITH, G. BERSUKER,
W. TAYLOR and R. JAMMY

SEMATECH, Austin, TX and Albany, NY, USA

We report on terahertz detection (from 0.2 THz to 2.4 THz) by Si FinFETs of different widths (with 2, 20, and 200 fins connected in parallel). FinFETs (with a small number of fins and with feature sizes as short as 20 nm to 40 nm) showed a very high responsivity (far above that previously measured for standard CMOS). We explain this improvement by negligible narrow channel effects.

Keywords: Terahertz; Detectors; Sensors; Plasma Wave Electronics.

1. Introduction

Within the past several years, the terahertz range of the electromagnetic spectrum has gained ever increasing attention as potential applications emerge in communications, materials identification and imaging [1]. The limited power of available terahertz sources significantly impacts detector performance requirements. In addition, many applications require compact size, fast response time and room temperature operation. Commercially available detectors include pyro-electric devices, Golay cells and Schottky diodes. Typical figures of merit for these devices are summarized in Table 1.

Emerging plasma wave THz and sub-THz detectors have an advantage in their extremely fast speed and are suitable for integration with conventional VLSI technologies. However, their application has been hampered by relatively low responsivity and high Noise Equivalent Power (NEP). Their responsivity varies in a very wide range (from 1 V/W to 1000 V/W depending upon materials system and device structure) and is especially affected by coupling of the THz radiation to the device. In devices without antenna structures, such coupling often involves the contact pads (including the gate pad) and bonding wires [2].

In our present research, we show that since coupling typically involves the gate bonding pad, the device responsivity is decreased by distributive effects of the induced THz current along the gate. Due to this effect only a section of the device close to the bonding pad participates in the THz detection, with the remaining device width effectively shunting the load and decreasing the response. The width of the active section at high frequencies is inversely proportional to the square root of frequency and at high frequencies (above one THz) might be as small as a fraction of a micron. At such small widths however, narrow channel effects might become detrimental to the performance of conventional devices. The FinFET geometry avoids these effects facilitating significant improvements in both responsivity and NEP. In agreement with our model, the decrease in responsivity with frequency of incident radiation becomes less pronounced as the number of device fins decreases.

Table 1 - Commercially Available Room Temperature THz Detectors

	NEP $W/Hz^{1/2}$	Responsivity V/W	Response Time (seconds)	Operating Temp (K)
Golay cells [3]	10^{-10}	10^{5}	10^{-2}	300
Pyroelectrics [4]	10^{-10}	10^{5}	10^{-2}	240-350
Schottky diodes [5]	10^{-12}	10^{3}	10^{-12}	10-420

2. Plasma Wave Detectors

In their seminal paper of 1993, Dyakonov and Shur proposed device operation based on electron plasma waves, localized time-varying perturbations in electron density within the FET channel [6]. This two dimensional electron gas was shown to obey the equations of hydrodynamic motion and continuity, and in their subsequent work on this topic [7, 8] they presented plasma wave devices functioning as detectors and emitters of terahertz radiation and as mixers and frequency multipliers in the terahertz range.

Dependent upon the incident radiation frequency, material momentum relaxation time and device dimension, both resonant and non-resonant detectors were considered. Resonant detection requires the criteria $\omega\tau >> 1$ and $s\tau / L >> 1$, where ω and τ are the angular frequency of the incident radiation and the momentum relaxation time respectively, s is the plasma wave velocity and L is the device channel length. The plasma wave velocity is shown to be dependent upon the gate bias as $s = (q\, V_{gt} /m)^{1/2}$, where q is the electronic charge, m is the electron effective mass and $V_{gt} = V_{gs} - V_t$ is the gate bias swing relative to the threshold voltage V_t. Thus such resonant detectors are tunable via the gate bias. For devices in which $\omega\tau >> 1$ but $s\tau /L << 1$, i.e. relatively longer gates or where $\omega\tau << 1$, i.e. rapid momentum relaxation, detection will be non-resonant, (broadband).

Terahertz detection in silicon devices is typically non-resonant, (although at gate lengths sufficiently short so as to result in ballistic carrier transport, resonant detection in silicon may be possible). The non-resonant DC THz response voltage, δv, of an ideal FET is given by [6]:

$$\delta v = \frac{q v_a^2}{4ms^2} \left\{ \frac{1}{1 + K \exp\left(-\frac{qV_{gt}}{\eta k_B T} \right)} - \frac{1}{\left[1 + K \exp\left(-\frac{qV_{gt}}{\eta k_B T} \right) \right]^2 \left(\sinh^2 \kappa + \cos^2 \kappa \right)} \right\} \tag{1}$$

where v_a is magnitude of the AC voltage induced between the gate and source terminals by the incident radiation and $\kappa = (L/s)(\omega/2\tau)^{1/2}$. Here, η is the device sub-threshold ideality factor, k_B is the Boltzmann constant and T is the temperature in degrees Kelvin. The parameter K (ref. [9] uses κ) resulting in response attenuation in the sub-threshold region is attributed to gate-to-channel leakage current and is calculated from the leakage current density j_0 as:

$$K = \frac{j_0 L^2 mq}{2C\tau(\eta k_B T)^2} \tag{2}$$

where C is the gate capacitance per unit area. The plasma wave velocity s is calculated from the surface carrier concentration n_s as:

$$s = \left(\frac{q}{m} \frac{n_s}{dn_s/dv} \right)^{1/2} \tag{3}$$

with:

$$n_s = n^* \ln\left[1 + \exp\left(\frac{qV_{gt}}{\eta k_B T} \right) \right] \tag{4}$$

where $n^* = C\eta k_B T/q^2$. Thus in the regions above and below the device threshold, $n = CV_{gt}/q$ and $n = n^* exp(qV_{gt}/\eta k_B T)$ respectively, and is interpolated in the region near the threshold.

In contemporary silicon MOSFETs, despite gate dielectric thicknesses on the order of only a few nanometers, gate leakage is often vanishingly small, therefore $K \rightarrow 0$. In the absence of gate leakage, Stillman, et al. [10] attribute response attenuation in the sub-threshold region to voltage division between the device channel resistance and the resistance of the load. With this in mind, eq. (1) may be rewritten as:

$$\delta v = \frac{v_a^2}{4ms^2} \left(\frac{\sinh^2 \kappa - \sin^2 \kappa}{\sinh^2 \kappa + \cos^2 \kappa} \right) \left(\frac{1}{1 + R_{CH}/R_L} \right) \tag{5}$$

where R_{CH} and R_L are the channel and load resistances, respectively.

Until this point, we have considered only the case where drain current is nearly zero, i.e. "open drain" detection. Lu and Shur first demonstrated the substantial increase in detector response with the application of drain bias [11]. Veksler et al. [12], expand upon this, calculating the response for short samples where $L << s/(\omega/\tau)^{1/2}$ as:

$$\delta v = -\frac{\lambda}{(1-\lambda)^{3/2}} \frac{v_a^2}{4V_{gt}} \tag{6}$$

For long samples where $L >> s/(\omega/\tau)^{1/2}$, the response is given as:

$$\delta v = \frac{v_a^2}{4V_{gt}} \frac{1}{\sqrt{1-\lambda}} \tag{7}$$

In both eqs. (6) and (7), $\lambda = j_d/j_{dsat}$ thus the response rises dramatically as the device is biased into saturation. (While these equations apply only above the device threshold, the more general case which applies to the sub-threshold region was also considered in [12]). While eqs. (5) - (7) accurately predict the response dependence on gate and drain bias, particularly near and below the device threshold and in the region at the onset of saturation, the responsivity of the device is strongly affected by coupling of the THz radiation to the device.

3. FinFET Structure

Dual gate Si MOSFETs were proposed in the early 1990s as a solution to the intractable problem of threshold voltage control as device gate lengths entered the sub-100 nm regime [13-15]. The conventional method of limiting short channel effects by channel doping required untenably high impurity concentrations (on the order of 10^{18} cm^{-3}) likely to degrade device performance through reductions in carrier field mobility and threshold voltage shifts due to quantization of carrier energies [16]. Initial schemes relied on a "vertical" architecture where back and front side gates were placed below and above the device channel. This approach represented significant challenges to existing semiconductor processing. Hisamoto et al. presented an alternative quasi-planar structure in 2000 where the gate conductor wraps around a vertically formed channel "fin"; constructed devices were found to exhibit avoidance of short channel effects with gate lengths as short as 17 nm [17]. The FinFET devices used in our experiments are of similar design, shown schematically in Figure 1, with 2, 20 or 200 fins of 40 nm height and designed widths from 40 to 100 nm. The designed gate length range was from 40 to 100 nm. Fin widths are typically reduced in processing on the order of 10 – 20 nm; gate lengths are expected to be approximately 5 nm shorter than as designed.

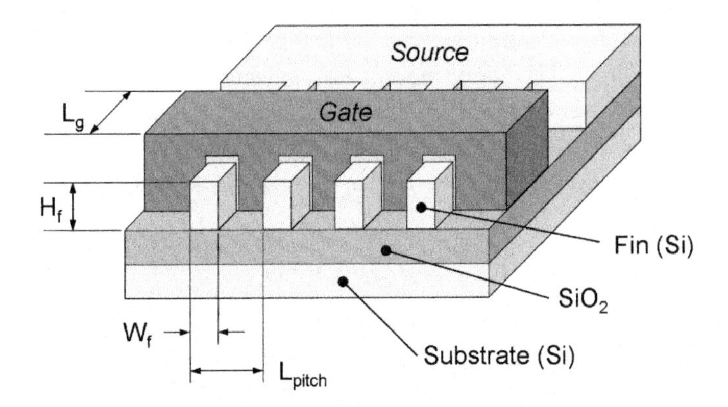

Figure 1 – Schematic illustration of FinFET device structure (the drain region is omitted from the foreground for clarity). The vertical channel fins are surrounded by the gate dielectric and gate conductor forming an effective dual gate MOSFET. Note that the dielectric along the top surfaces of the fin may be increased in order to reduce parasitic capacitance.

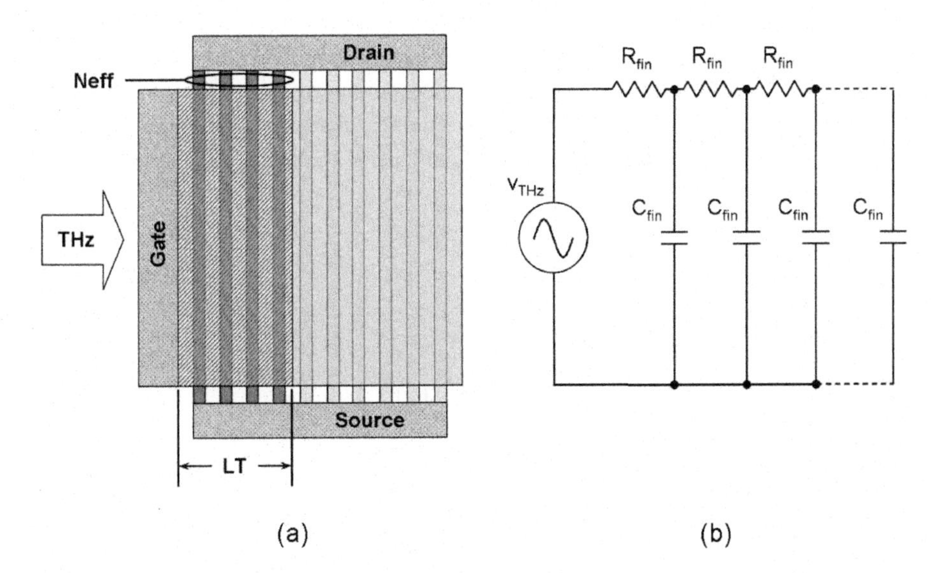

Figure 2 – (a) Conceptual illustration of the effective number of fins (in red) in a given device contributing to response. Fins beyond the characteristic transfer length L_T behave as a load to the response. (b) Schematic representation of equivalent circuit at Terahertz frequencies.

In order to model the frequency dependence of the active area of the FinFET structure we conceptualize the device as shown in Figure 2(a). Here the incident radiation is coupled into the device from the bond pad connection and the signal propagation along the nearly lossless gate conductor transmission line across the device. Figure 2(b) shows the equivalent circuit at THz frequencies. The inductance of the gate conductor is assumed to be greatly overshadowed by that of the bond wires leading to the device and is therefore neglected. Similarly, the conductance of the gate dielectric is neglected since gate leakage is vanishingly small. The fin capacitance, C_{FIN}, and fin access resistance, R_{FIN}, are calculated as:

$$C_{FIN} = \frac{\varepsilon\left(2H_f + W_f\right)L_g}{d} \tag{8}$$

and:

$$R_{FIN} = \frac{\rho_g L_{PITCH}}{L_g} \tag{9}$$

where ε and d are the gate dielectric permittivity and thickness, respectively, L_g is the gate length, H_f and W_f are the fin height and width respectively, L_{PITCH} is the fin pitch and ρ_g is the resistivity of the gate conductor. At some multiple of the characteristic transfer length, L_T, the signal is attenuated so as to be negligible. Only the fins prior to this point, N_{eff}, are effective in contributing to the device response; the remaining fins serve only as an additional load. For a large number of fins, the voltage and current distributions at THz frequency along the gate can be using the telegrapher's equations:

$$\frac{\partial^2 V(x)}{\partial x^2} = \Gamma^2 V(x) \tag{10}$$

and:

$$\frac{\partial^2 I(x)}{\partial x^2} = \Gamma^2 I(x) \tag{11}$$

where $V(x)$ and $I(x)$ are the voltage and current signals along the line respectively, and:

$$\Gamma = \sqrt{j\omega RC} \tag{12}$$

Here ω is the angular frequency of the incident radiation and R and C are the per unit length values of the effective resistance of the gate conductor and the gate capacitance, respectively, calculated as:

$$R = R_{FIN} / L_{PITCH} \tag{13}$$

and:

$$C = C_{FIN} / L_{PITCH} \tag{14}$$

The characteristic impedance of the transmission line is then:

$$Z_0 = \sqrt{\frac{R}{j\omega C}} \tag{15}$$

In the case of reasonably small R, the signal is found to decay as:

$$V(x) = \mathrm{e}^{-x/L_T} \tag{16}$$

with L_T as given by:

$$|L_T| = \frac{2Z_0}{R} = \frac{2}{\sqrt{\omega RC}} \tag{17}$$

To calculate N_{eff}, L_T is normalized to the fin pitch L_{PITCH} (see Figure 1). While the incident signal is attenuated exponentially along the transmission length, for simplicity we consider only those fins at a distance closer to the gate pad than L_T as contributing to the response and those beyond as passive loads, thus N_{eff} is given by:

$$N_{eff} = \min(N, N_{LT}) \tag{18}$$

where:

$$N_{LT} = \frac{L_T}{L_{PITCH}} \tag{19}$$

With this expression for N_{eff}, we may calculate the active portion of the multi-channel FET channel resistance (R_A) as:

$$R_A = \frac{1}{N_{eff}}\left[R_S + R_D + \frac{L_g}{(2H_f + W_f)q\mu n_s}\right] \tag{20}$$

When $N \leq N_{eff}$, all device fins are active; when $N > N_{eff}$, the resistance of the passive fins is found as:

$$R_P = \frac{1}{(N - N_{eff})}\left[R_S + R_D + \frac{L_g}{(2H_f + W_f)q\mu n_s}\right] \tag{21}$$

We now consider the effective circuit for the DC THz response as shown in Figure 3(a), and the Thevenin equivalent circuit in Figure 3(b), where:

$$V_{Th} = \delta v\left(\frac{R_P}{R_P + R_A}\right) \tag{22}$$

and:

$$R_{Th} = \frac{R_A R_P}{R_A + R_P} \tag{23}$$

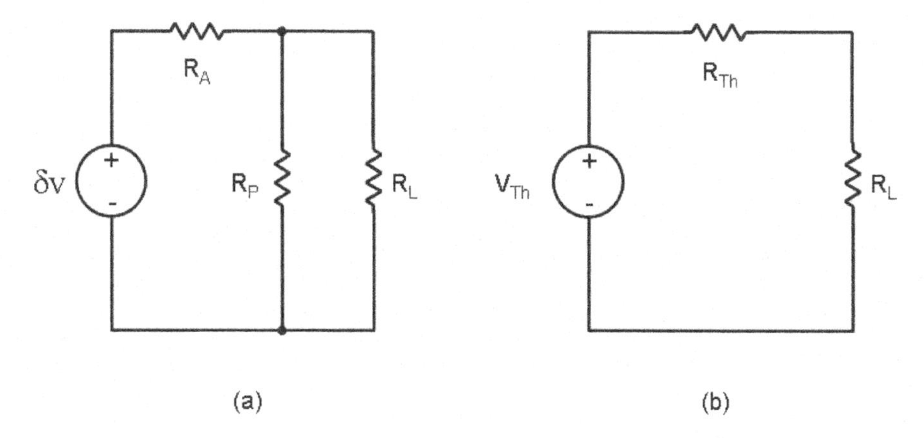

Figure 3 – (a) DC equivalent circuit for FinFET, and (b) Thevenin equivalent representation.

Hence, the detector response is now given by:

$$\delta v_{eff} = \frac{v_a^2}{4ms^2}\left(\frac{\sinh^2\kappa - \sin^2\kappa}{\sinh^2\kappa + \cos^2\kappa}\right)\frac{R_P R_L}{R_L\left(R_A + R_P\right) + R_P R_A} \tag{24}$$

Until this point, we have ignored the effects of coupling of the incident radiation to the device. While there certainly are a great many factors to consider, for our purpose of modeling the frequency response of the device without an optimized antenna, we consider only the additional effects of the bond wire inductance on the response. We model the attenuation of the incident signal using a lumped parameter voltage divider between the device capacitance and series connected device resistance and bond wire inductance, resulting in the following expression for the effective coupled radiation:

$$v_a = v_{THz}\frac{Z_o}{Z_o + Z_{THz}} \tag{25}$$

Where Z_o is given by eq. (15). We assume that the impedance of the effective THz source is inductive with the effective inductive impedance given by:

$$Z_{THz} = j\omega L_{THz} \tag{26}$$

Combining eqs. (24) and (25) our response expression becomes:

$$\delta v_{eff} = \frac{v_{THz}^2}{4ms^2}\left(\frac{\sinh^2\kappa - \sin^2\kappa}{\sinh^2\kappa + \cos^2\kappa}\right)\left(\frac{R_P R_L}{R_L\left(R_A + R_P\right) + R_P R_A}\right)\left(\frac{Z_o}{Z_o + Z_{THz}}\right)^2 \tag{27}$$

It is expected that this expression will be applicable to the drain current enhanced response of eqs. (6) and (7) with the appropriate adjustments as discussed previously.

The speed of detector response is also of importance. Kachorovskii and Shur [18] propose theoretical calculations of the maximum response modulation frequency in plasma wave detectors as shown in the following expressions:

$$f_{\max} = \begin{cases} \dfrac{\mu_f V_{gt}}{2\pi L_g^2} & V_{gt} > 0,\ qV_{gt} >> T \\ \\ \dfrac{\mu_f \eta k_B T}{2\pi q L_g^2} & V_{gt} < 0,\ q\left|V_{gt}\right| >> T \end{cases} \tag{28}$$

Here μ_f is the effective field effect mobility, which is very different from conventional mobility for short channel devices, where ballistic or near ballistic transport is dominant. Near $V_{gt} = 0$, f_{max} is interpolated as:

$$f_{\max} = \left(\frac{\mu_f \eta k_B T}{2\pi q L_g^2}\right)\left[1 + \exp\left(\frac{-qV_{gt}}{\eta k_B T}\right)\right]\ln\left[1 + \exp\left(\frac{qV_{gt}}{\eta k_B T}\right)\right] \tag{29}$$

4. Response Measurements

Prior to response measurements, current / voltage characteristics were performed and die containing several functional devices were wire bonded within ceramic chip carriers to simplify handling. Response measurements were made using a standard lock-in technique. Two radiation source types were used: a purpose built Gunn diode oscillator equipped with frequency multipliers was used at the 0.2 and 0.6 THz. An optically pumped terahertz gas laser was used at 1.6 and 2.4 THz. Source power was measured as 1.5 mW, 35 μW, 30 mW and 10 mW for 0.2, 0.6, 1.6 and 2.4 THz respectively. A chopper was placed in the beam path and the radiation was focused onto the device using either a parabolic mirror or polyethylene lens depending upon the source. Two programmable power supplies were used to provide gate and drain bias. Figure 4 illustrates the typical response dependence upon gate bias.

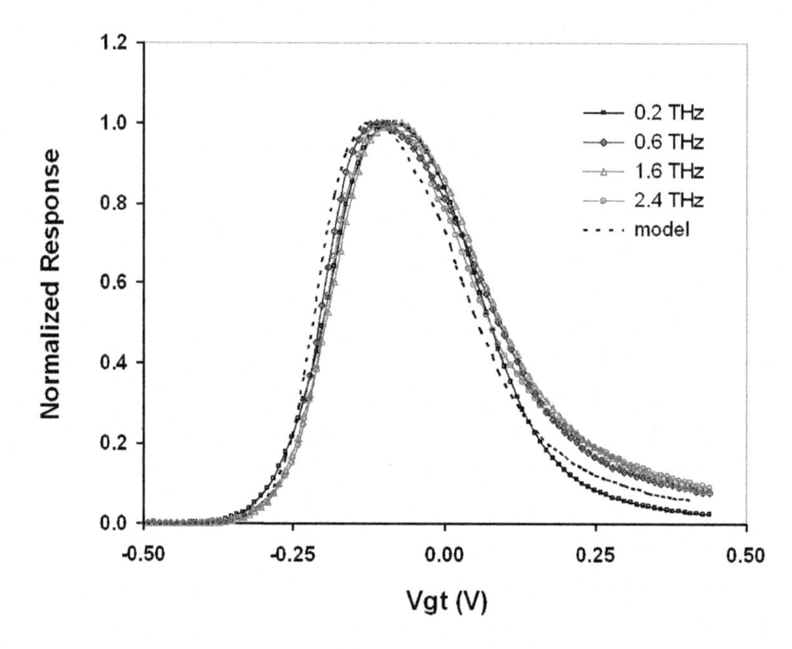

Figure 4 – 20 fin nFET open drain response, normalized to response maxima, across several incident frequencies. $W_f = 40$ nm; $L_g = 100$ nm. Dashed line represents modeled response using eq. (5) with $f = 0.2$ THz.

The response at each frequency is normalized to its peak value to allow direct comparison across the frequency range, and good agreement with the model of eq. (5) is apparent. That the response peaks coincide across the frequency range confirms the non-resonant nature of the response. Figure 5 compares responsivity modeled using eq. (27) to measured data for several FinFET devices. (Calculations of responsivity in measured data are simply measured response divided by incident power; no adjustment for device vs. beam size is used since it is not clear that this approach is warranted [19]). Note that while there are anomalies, the fit is reasonable, especially in the response attenuation at and above 0.6 THz. In addition, note that the FinFET devices in many cases exhibit considerably greater responsivity than standard CMOS FETs.

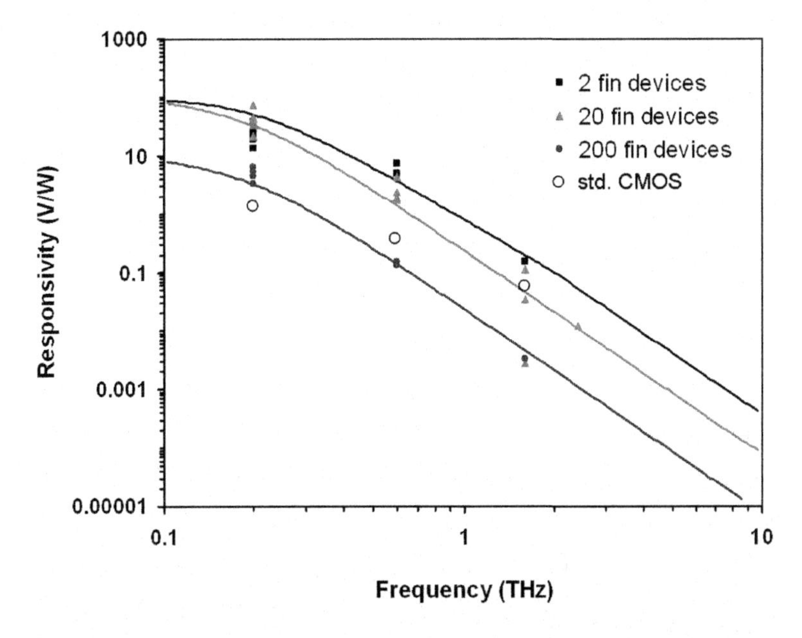

Figure 5 – Open drain responsivity of several FinFET devices at various incident frequencies. Filled symbols are measured data; lines are modeled response using eq. (27) with L_g = 100 nm, L_{pitch} = 200 nm, H_f = 40 nm, W_f = 60 nm, $\varepsilon/\delta = 6.9\text{x}10^{-2}$ F/m^2, ρ_g = 10 ohms/□, $R_S = R_D$ = 300 ohms, μ = 0.5 m^2/Vs, V_{gt} = -0.1 V. V_{THz} and L_{THz} were chosen to fit measured data. Open symbols are measured responsivity for standard CMOS FETs. [19, 20]

Responsivity rises dramatically as expected in FinFET devices, reaching from several hundred to above one thousand volts per watt. Figure 6 illustrates the degree to which the enhanced responsivity follows that predicted with eq. (6). In Figure 7 is shown the drain current enhanced responsivity for several FinFETs at various frequencies compared with the response of standard CMOS FET data. Here especially is demonstrated the advantage of the FinFET structure, as responsivity is seen to be nearly two orders of magnitude higher than the standard CMOS FETs.

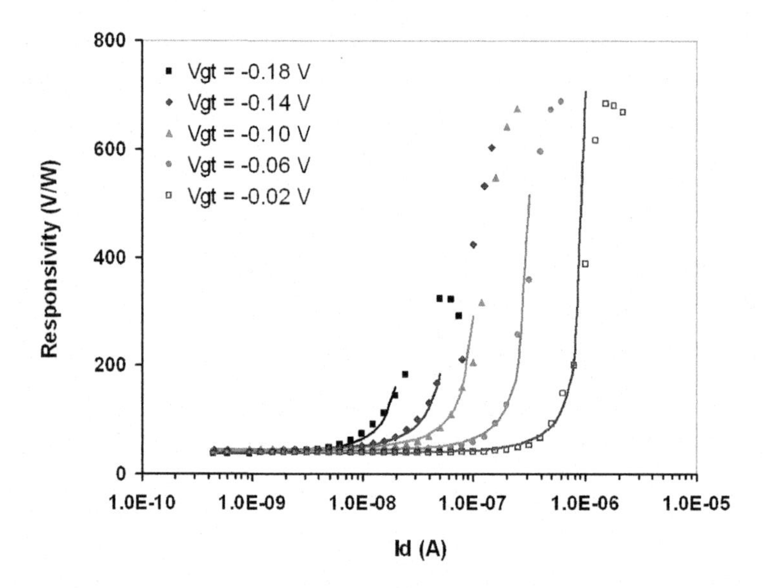

Figure 6 – 0.2 THz responsivity of 20 fin device with $W_f = 40$ nm and $L_g = 100$ nm. Symbols are measured data; lines are modeled responsivity following eq. (6).

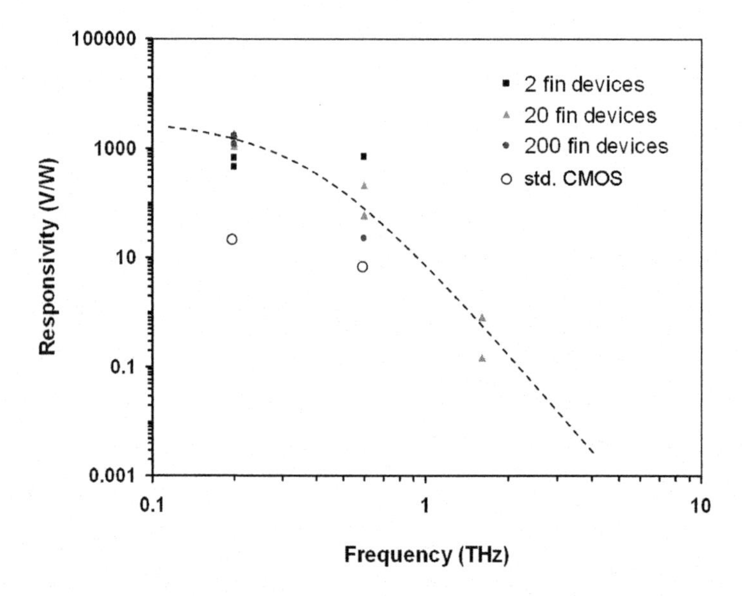

Figure 7 – Peak drain current enhanced responsivity of several FinFET devices vs. incident frequency. Filled symbols are measured data. Dashed line is drawn to guide the eye. Open symbols are measured responsivity for standard CMOS FETs. [20]

5. Noise Equivalent Power

Noise Equivalent Power represents the minimum signal power distinguishable from the detector noise, and is a figure of merit for the detector sensitivity. Commonly, NEP is considered to be the minimum power detectable per square root of bandwidth, in units of W/√Hz, and is calculated as the inverse ratio of responsivity to the square root of device voltage noise. In the open drain configuration, the thermal noise of the channel resistance is predominant [10, 21]. Reducing the channel resistance decreases NEP for a given responsivity by a factor of the square root of the resistance reduction, thus increasing the number of device fins from 2 to 20, or 20 to 200 will decrease the noise contribution to NEP by a factor of ~3.2. Returning to Figure 5 however, finds a decrease in responsivity of one magnitude order between 20 and 200 fin devices, therefore NEP increases for the larger number of fins. This is illustrated in Figure 8.

The comparison between 2 and 20 fins is less clear due to the observed anomalies in responsivity for the 2 fin devices, though in general NEP is expected again to be higher for the larger number of fins.

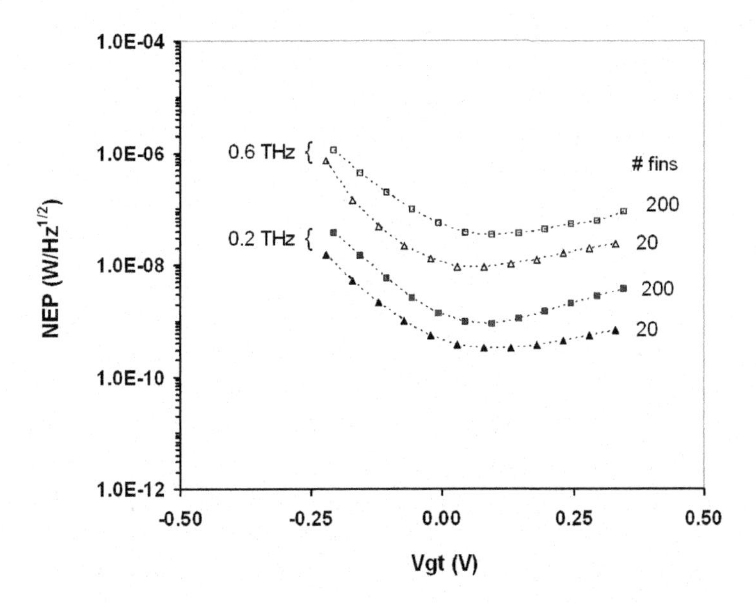

Figure 8 – Noise equivalent power vs. gate bias for two FinFET devices at 0.2 (filled symbols) and 0.6 THz (open symbols) calculated from measured response and channel resistance data. $W_f = 40$ nm, $L_g = 60$ nm, 20 fins (triangles) and 200 fins (squares).

Estimation of NEP becomes more complicated with the introduction of drain current. Responsivity increases on average between 20 and 30 times that of the open drain response for the FinFET devices; however device noise increases dramatically as well, essentially as a function of the drain current squared. The noise spectra are seen to follow a $1/f^a$ distribution, with $a \approx 1$, as is shown in Figure 9. Here the spectrum measured with a drain load resistance of 2.5 kohms is used to estimate the spectra with loads more typical of those used in responsivity measurements, allowing comparison of NEP with drain current to that of the open drain configuration.

Peak responsivity for the device in Figure 9 was measured to be 1200 V/W; at the optical chopper frequency used in response measurements of 50 Hz and with an amplifier load of approximately 10 Mohm, the voltage noise density is calculated to be 5 x 10^{-8} V^2/Hz, thus the NEP at 50 Hz is approximately 2 x 10^{-7} W/Hz$^{1/2}$. The sampling frequency required to achieve the open drain NEP for this device of 3 x 10^{-3} W/Hz$^{1/2}$ is calculated to be approximately 35 MHz, well below the ~30 GHz maximum response frequency predicted by eq. (29).

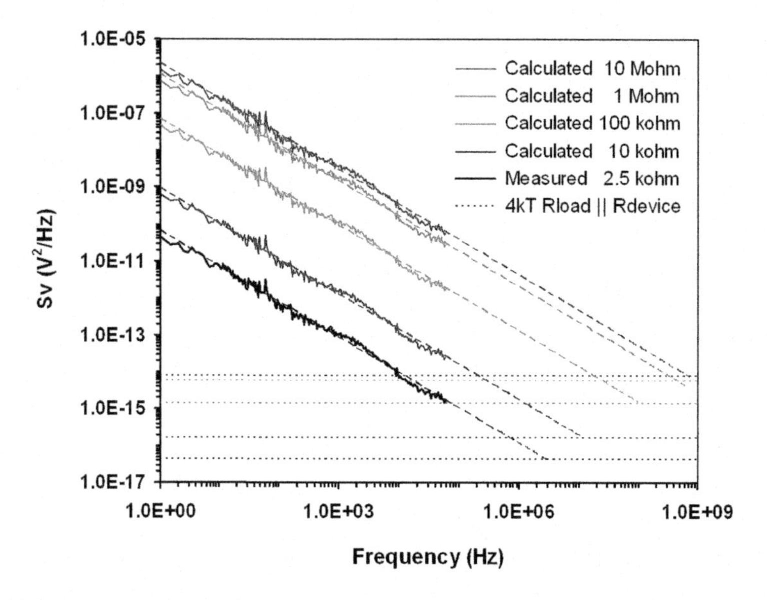

Figure 9 – Voltage noise spectral density for 20 fin device with W_f = 40 nm and L_g = 100 nm. Measured data at V_{gt} = 0V and I_d = 3.1 μA with 2.5 kohm drain load (lowest line) is calculated to show expected device noise for larger values of load resistance (responsivity measurements presented earlier are loaded at 10 Mohm). Horizontal dashed lines are corresponding calculated thermal noise values; angled dashed lines project device noise to these values to indicate frequencies of equivalence.

6. Conclusion

We have demonstrated the response of Si FinFETs to terahertz and sub-terahertz radiation and developed a model to explain the observed decrease in response as the number of device fins increases, which we attribute to the attenuation of the coupled incident radiation both across the device gate conductor transmission line and due to the inductance of the device bond wires. Our results show that narrow FinFETs can be competitive or better than commercial THz detectors potentially enabling the development of sub-THz and THz cameras implemented using this technology.

Acknowledgments

The work at RPI is supported by SEMATECH, by the InterconnectFocusCenter, and by the NSF under the auspices of the I/UCRC "CONNECTION ONE".

References

[1] W. J. Stillman and M. S. Shur, "Closing the Gap: Plasma Wave Electronic Terahertz Detectors," *Journal of Nanoelectronics and Optoelectronics*, vol. 2, pp. 209-221, 2007.
[2] T. A. Elkhatib, A. V. Muravjov, D. B. Veksler, W. J. Stillman, X.-C. Zhang, M. S. Shur, and V. Y. Kachorovskii, "Subwavelength Detection of Terahertz Radiation using GaAs HEMTs," Proceedings of IEEE Sensors, pp. 1988-1990, 2009.
[3] QMC_Instruments, http://www.terahertz.co.uk/,2010
[4] Spectrum_Detector_Inc., http://www.spectrumdetector.com/pdf/datasheets/THZ.pdf,2007
[5] Virginia_Diodes_Inc,http://virginiadiodes.com/WR2.2ZBD.htm,2007
[6] M. Dyakonov and M. Shur, "Shallow water analogy for a ballistic field effect transistor: New mechanism of plasma wave generation by dc current," *Physical Review Letters*, vol. 71, pp. 2465-2468, 1993.
[7] M. I. Dyakonov and M. S. Shur, "Plasma wave electronics: Novel terahertz devices using two dimensional electron fluid," *IEEE Transactions on Electron Devices*, vol. 43, pp. 1640-1645, 1996.
[8] M. Dyakonov and M. Shur, "Detection, mixing, and frequency multiplication of terahertz radiation by two-dimensional electronic fluid," *IEEE Transactions on Electron Devices*, vol. 43, pp. 380-387, 1996.
[9] W. Knap, V. Kachorovskii, Y. Deng, S. Rumyantsev, J. Q. Lu, R. Gaska, M. S. Shur, G. Simin, X. Hu, M. Asif Khan, C. A. Saylor, and L. C. Brunel, "Nonresonant detection of terahertz radiation in field effect transistors," *Journal of Applied Physics*, vol. 91, pp. 9346, 2002.
[10] W. Stillman, M. S. Shur, D. Veksler, S. Rumyantsev, and F. Guarin, "Device loading effects on nonresonant detection of terahertz radiation by silicon MOSFETs," *Electronics Letters*, vol. 43, pp. 422-423, 2007.
[11] J. Q. Lu and M. S. Shur, "Terahertz detection by high-electron-mobility transistor: Enhancement by drain bias," *Applied Physics Letters*, vol. 78, pp. 2587, 2001.
[12] D. Veksler, F. Teppe, A. P. Dmitriev, V. Y. Kachorovskii, W. Knap, and M. S. Shur, "Detection of terahertz radiation in gated two-dimensional structures governed by dc current," *Physical Review B (Condensed Matter and Materials Physics)*, vol. 73, pp. 125328-1, 2006.
[13] M. Schubert, B. Hofflinger, and R. P. Zingg, "A one-dimensional analytical model for the dual-gate-controlled thin-film SOI MOSFET," *Electron device letters*, vol. 12, pp. 489-491, 1991.
[14] D. J. Frank, S. E. Laux, and M. V. Fischetti, "Monte Carlo simulation of a 30 nm dual-gate MOSFET: how short can Si go?," International Electron Devices Meeting 1992. Technical Digest (Cat. No.92CH3211-0), pp. 553-6, 1992.
[15] K. Suzuki, T. Tanaka, Y. Tosaka, H. Horie, and Y. Arimoto, "Scaling theory for double-gate SOI MOSFET's," *IEEE Transactions on Electron Devices*, vol. 40, pp. 2326-9, 1993.

[16] C. Fiegna, H. Iwai, T. Wada, M. Saito, E. Sangiorgi, and B. Ricco, "Scaling the MOS transistor below 0.1 m: methodology, device structures, and technology requirements," *IEEE Transactions on Electron Devices*, vol. 41, pp. 941-51, 1994.
[17] D. Hisamoto, L. Wen-Chin, J. Kedzierski, H. Takeuchi, K. Asano, C. Kuo, E. Anderson, K. Tsu-Jae, J. Bokor, and H. Chenming, "FinFET-a self-aligned double-gate MOSFET scalable to 20 nm," *IEEE Transactions on Electron Devices*, vol. 47, pp. 2320-5, 2000.
[18] V. Y. Kachorovskii and M. S. Shur, "Field effect transistor as ultrafast detector of modulated terahertz radiation," *Solid State Electronics*, vol. 52, pp. 182-5, 2008.
[19] W. J. Stillman, "Silicon CMOS FETs as terahertz and sub-terahertz detectors," in *Doctoral Thesis*. Troy, NY: RPI, 2008.
[20] W. Stillman, F. Guarin, V. Y. Kachorovskii, N. Pala, S. Rumyantsev, M. S. Shur, and D. Veksler, "Nanometer scale complementary silicon MOSFETs as detectors of terahertz and sub-terahertz radiation," IEEE Sensors 2007 Conference, pp. 934-7, 2007.
[21] R. Tauk, F. Teppe, S. Boubanga, D. Coquillat, W. Knap, Y. M. Meziani, C. Gallon, F. Boeuf, T. Skotnicki, C. Fenouillet-Beranger, D. K. Maude, S. Rumyantsev, and M. S. Shur, "Plasma wave detection of terahertz radiation by silicon field effects transistors: Responsivity and noise equivalent power," *Applied Physics Letters*, vol. 89, pp. 253511, 2006.

PROGRESS IN DEVELOPMENT OF ROOM TEMPERATURE CW GASB BASED DIODE LASERS FOR 2-3.5 μM SPECTRAL REGION

TAKASHI HOSODA

Department of Electrical and Computer Engineering, State University of New York at Stony Brook, NY 11794, USA
thosoda@ece.sunysb.edu

JIANFENG CHEN

Department of Electrical and Computer Engineering, State University of New York at Stony Brook, NY 11794, USA
cjf2006510@gmail.com

GENE TSVID

Department of Electrical and Computer Engineering, State University of New York at Stony Brook, NY 11794, USA
evgeny.tsvid@stonybrook.edu

DAVID WESTERFELD

Power Photonic Corporation, Stony Brook, New York 11790, USA
davidwesterfeld@yahoo.com

RUI LIANG

Department of Electrical and Computer Engineering, State University of New York at Stony Brook, NY 11794, USA
liangr85@gmail.com

GELA KIPSHIDZE

Department of Electrical and Computer Engineering, State University of New York at Stony Brook, NY 11794, USA, Power Photonic Corporation, Stony Brook, New York 11790, USA
gela@ece.sunysb.edu

LEON SHTERENGAS

Department of Electrical and Computer Engineering, State University of New York at Stony Brook, NY 11794, USA
leon@ece.sunysb.edu

GREGORY BELENKY

Department of Electrical and Computer Engineering, State University of New York at Stony Brook, NY 11794, USA, Power Photonic Corporation, Stony Brook, New York 11790, USA
garik@ece.sunysb.edu

Recent progress and state of GaSb based type-I lasers emitting in spectral range from 2 to 3.5 μm is reviewed. For lasers emitting near 2 μm an optimization of waveguide core width and asymmetry allowed reduction of far field divergence angle down to 40-50 degrees which is important for improving coupling efficiency to optical fiber. As emission wavelength increases laser characteristics degrade due to insufficient hole confinement, increased Auger recombination and deteriorated transport through the waveguide layer. While Auger recombination is thought to be an ultimate limiting factor to the performance of these narrow bandgap interband lasers we demonstrate that continuous improvements in laser characteristics are still possible by increasing hole confinement and optimizing transport properties of the waveguide layer. We achieved 190, 170 and 50 mW of maximum CW power at 3.1, 3.2 and 3.32 μm wavelengths respectively. These are the highest CW powers reported to date in this spectral range and constitute 2.5-fold improvement compared to previously reported devices.

Keywords: GaSb; InGaAsSb; type-I; diode lasers; quantum well lasers; midinfrared; long wavelength.

1. Introduction

Semiconductor lasers emitting in 2 to 3.5 μm wavelength range are in demand for laser spectroscopy, medical diagnostics, medical therapy and solid state laser pumping [1-5]. This demand is being met by developing GaSb based type-I diode laser technology. At the short wavelength end of the region InGaSb quantum well lasers on GaSb substrate show excellent performance. High power 2 μm diode lasers with broadened waveguide achieved threshold current densities of 100 A/cm2. Differential quantum efficiencies of 53% for 2-mm-long multimode devices were demonstrated [6-7]. Unfortunately broadened waveguide lasers have rather large fast axis divergence, typically about 65 degrees full width at half maximum (FWHM). To facilitate coupling of light to an optical fiber a divergence of 40 to 50 degrees is desirable. To this end we present an optimization of waveguide core width and asymmetry for 2 μm emitting devices. By adjusting waveguide and cladding parameters the fast axis divergence of 45-50 degrees and 1 W output power with 15% power conversion efficiency can be achieved. In the second part of the paper we review CW power performance over spectral range of 2 to 3.4 μm. Until recently the prospect of making diode lasers operating in CW at RT with wavelength near 3 μm was thought to be questionable. This pessimistic scenario originated from recognition of the well known fundamental increase of the probability of the nonradiative Auger recombination and free carrier absorption with wavelength. The associated carrier and photon losses were considered to prevent the mid-infrared diode lasers from reaching CW threshold at RT. However we demonstrate that optimization of waveguide composition and thickness results in improved performance. In particular devices with 32% of indium in 640 nm wide quinary InAlGaAsSb waveguide produced maximum CW powers of 190, 170 and 50 mW at 3.1, 3.2 and 3.32 μm wavelengths respectively.

2. Results and Discussion

2.1. *Waveguide core width and asymmetry optimization for 2 μm emitting devices*

The laser heterostructures were grown by solid-source molecular beam epitaxy in a Veeco GEN-930 reactor equipped with As and Sb valved cracker sources on GaSb

Te-doped substrates. Te and Be were used for n and p doping respectively. The laser active region comprised two 12 nm wide $In_{0.2}Ga_{0.8}Sb$ quantum wells separated by 20 nm of AlGaAsSb quaternary alloy lattice matched to GaSb substrate. The devices were nominally identical except for variations in waveguide core and cladding widths and aluminum content of n-cladding as shown in Table 1.

Table 1. Aluminum composition and width of the claddings and core layers of 2 μm emitting lasers.

Sample ID	p-cladding		Core		n-cladding	
	% Al	Thickness [nm]	% Al	Thickness [nm]	% Al	Thickness [nm]
1	90	1500	25	859	90	1500
2	90	1500	25	850	50	2000
3	90	1500	25	550	50	2000
4	90	1500	25	350	50	2000
5	90	1500	25	250	50	2500
6	90	1000	30	450	50	2500

Two parameters varied in this series are the waveguide core width and asymmetry of the mode field relative to N and P claddings. As expected the narrower waveguide core the larger is the mode field waist and hence smaller far field divergence angle, Figure 1f.

The penalty for larger penetration of the mode field into the claddings is an increase in internal losses. Since free hole absorption is expected to be higher than that of free electrons [8-9] it is beneficial to design asymmetric mode that penetrates into the n-cladding more than into the p-cladding. This can be done by lowering aluminum content of the n-cladding layer, increasing n-cladding thickness and decreasing p-cladding thickness. Continuous wave power characteristics of selective structures are given in Figure 1. Diode lasers with structure 3 (far field divergence with FWHM of 50 degrees) generate above 1.5 W of CW power and demonstrate peak power conversion efficiency in excess of 28%, Figure 1a. Linear laser array with fill factor of 20% composed of eighteen 2-mm-long emitters generated above 11 W at 20C at current of 55 A, Figure 1b. Power conversion efficiency was above 20% in peak and above 15% at maximum power level. Driving current and output power were limited by TEC stage heat removal capabilities. Diode lasers with structure 4 (far field divergence with FWHM of 44 degrees) generate 1.4 W of CW power but peak power conversion efficiency was below 20%, Figure 1c. Smaller value of the output power and efficiency is associated with higher optical loss and stronger temperature sensitivity of the threshold. Still the devices were producing more than watt of 2 μm power at power conversion efficiency better than 10%. Seven 2-mm-long AR/HR coated single emitters of structure 4 were epi-down mounted onto Au-coated BeO blocks and connected in series.

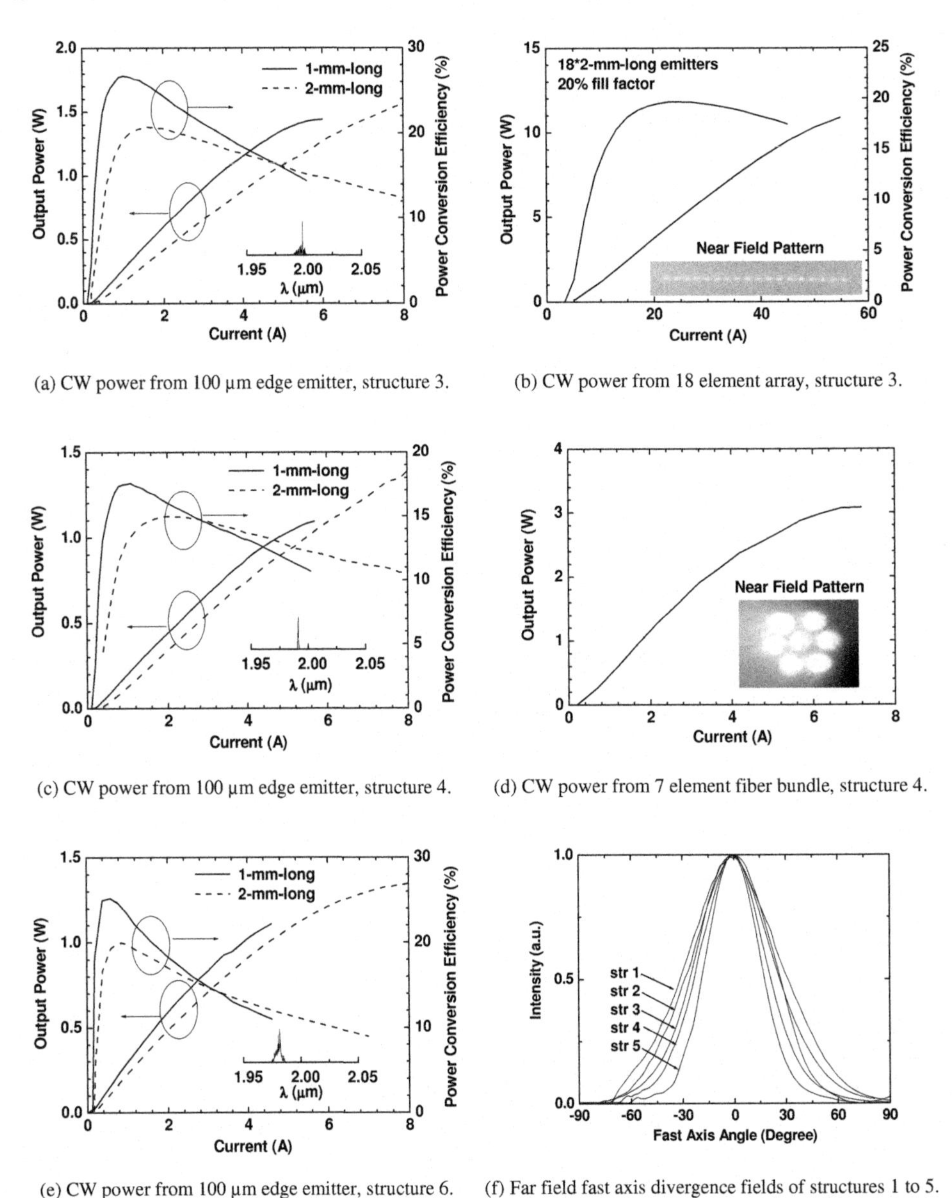

(a) CW power from 100 μm edge emitter, structure 3.

(b) CW power from 18 element array, structure 3.

(c) CW power from 100 μm edge emitter, structure 4.

(d) CW power from 7 element fiber bundle, structure 4.

(e) CW power from 100 μm edge emitter, structure 6.

(f) Far field fast axis divergence fields of structures 1 to 5.

Figure 1. Power outputs and power conversion efficiencies subfigures (a,c,e) of structures (3,4,6) respectively; array CW power outputs of structures 3,4, subfigures (b,d); fast axis far fields for structures 1 to 5 (f).

Output light was collected by a cylindrical micro-lens in front of each emitter into seven multimode silica fibers (core diameter of 105 μm). No optimization of the collecting optics was done for the particular 2 μm laser technology. More than 3 W of CW power was collected from the other end of the fiber bundle, Figure 1d. For structure 6 QWs were shifted closer to the n cladding layer. As low as 100 A/cm^2 and 85 A/cm^2 of CW threshold current density, at 20C, were achieved for 1 and 2-mm-long AR/HR coated lasers, respectively, Figure 1e. Output power above 1.3 W was generated at 8 A. Power conversion efficiency peaked above 25% but rapidly decreased with current due to voltage drop across series resistance. Increase of the series resistance in structure 6 as compared to structure 3, for instance, is not fully understood. The excessive voltage drop across laser heterostructure can be ascribed to reduced doping in p-cladding layer (10^{17} cm^{-3} of Be was used in p-cladding of structure 6 as compared to $2*10^{17}$ cm^{-3} in structures 1-4). This model would have to assume rather low hole mobility. Another possible explanation is an adverse variation of the substrate doping level and associated increase of the n-contact resistance.

2.2. *CW power characteristics over 2.2 to 3.3 μm wavelength range*

By increasing indium composition in the quantum well and adding arsenic as necessary to keep strain within permissible 2% level it is possible to extend emission wavelength up to 3.4 micron. Figure 2 shows CW power characteristics achievable at 2.2, 2.7, 3.0, 3.1, 3.2 and 3.32 μm emission wavelengths. Maximum CW powers are approximately 1500, 600, 300, 190, 170 and 50 mW. This degradation is due to an increase in Auger recombination rate [10], decrease in hole confinement [11] and, as more recently discussed, carrier recombination in the waveguide [12]. At 2.2 μm CW power of 1 Watt can be achieved with 17% wall-plug efficiency while at 2.7 μm we get 400 mW with 9% of power conversion efficiency. Recently it was found that two improvements can be made to devices emitting above 3 μm. First, due to carrier recombination in the waveguide narrower cores are preferred [12]. Second, hole confinement can be improved by increasing indium composition in the waveguide [13]. Application of these two concepts to lasers emitting above 3 μm resulted in CW characteristics shown in Figure 2b, [14]. Devices differ in indium composition in the quantum wells to achieve emission at 3.0, 3.1, 3.2 (3.2A and 3.2B) and 3.32 μm. Lasers 3.2A and 3.2B, both emitting near 3.2 μm, had indium concentrations of 25% and 32% in quinary AlInGaAsSb waveguide layer respectively. The laser 3.2B with 32% of indium in the waveguide produced 1.5 times more maximum CW power than the laser with 25% of indium, 3.2A, as we believe, due to increased hole confinement.

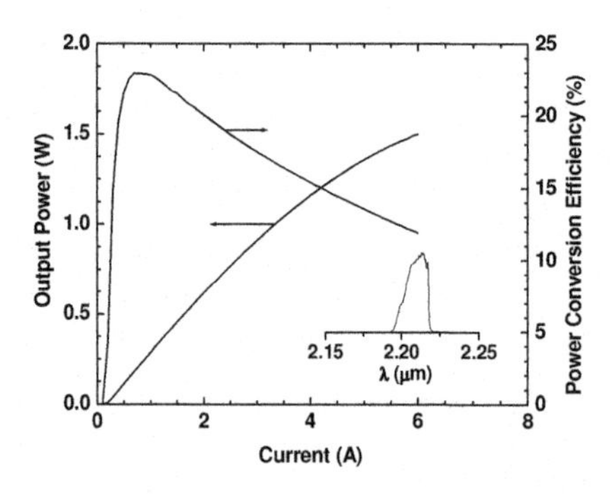

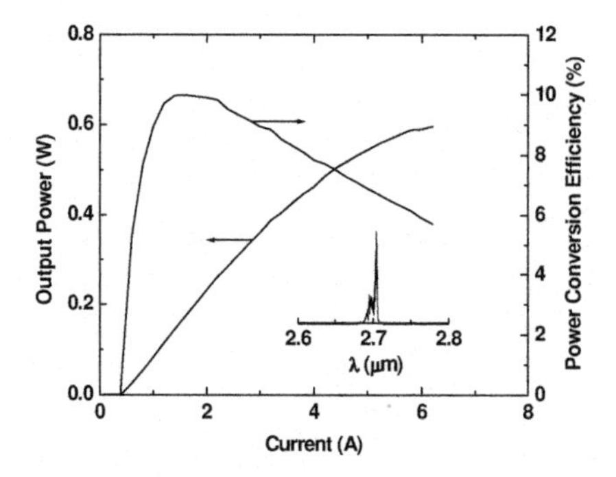

A) CW power (left axis) and power conversion efficiency (right axis) for lasers emitting at 2.2 and 2.7 µm. Insets are emission spectra.

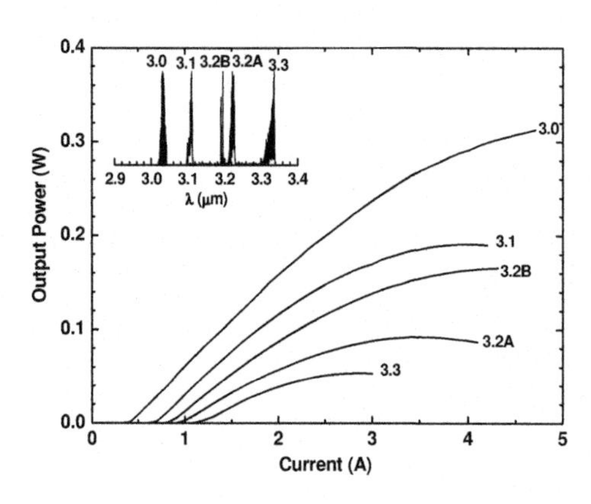

B) CW powers of 3.0, 3.1, 3.2 (3.2A- $Al_{0.22}In_{0.25}Ga_{0.53}AsSb$ and 3.2B- $Al_{0.22}In_{0.32}Ga_{0.46}AsSb$ lattice matched to GaSb waveguide layer) and 3.32 µm emitting 2-mm-long AR/HR coated lasers. The inset shows the laser spectra.

Figure 2. CW power performance of lasers in spectral region 2.2 - 3.3 µm. Inset shows emission spectra.

3. Conclusion

We reviewed design strategies and characteristics of the best InGaAsSb quantum well lasers emitting in 2 to 3.3 µm wavelength range. For 2 µm emitting lasers fast axis far field divergence angles of 45 to 50 degrees are attainable with only minor sacrifice to CW power and power conversion efficiency. Corresponding single emitters with 100 µm wide output aperture generate above 1 W at 20 C of 2 µm optical power with power conversion efficiency exceeding 15% at maximum power level and peaking at 28%. Linear laser arrays generate above 10 W of CW power. More than 3 W of CW power was demonstrated from fiber bundle containing seven 105-µm diameter silica fibers. Significant improvement to devices emitting above 3 µm was demonstrated. In particular lasers with 640 nm thick, 32% indium, quinary AlInGaAsSb waveguide layer have produced maximum CW powers of 300, 190, 170 and 50 mW for 3.0, 3.1 3.2 and 3.32 µm emission respectively. The improvement is attributed to increase in hole confinement and optimal width of the waveguide.

Acknowledgements

The work was supported by Air Force Office of Scientific Research by Grant FA95500810458 and Young Investigator Award FA95500810083, National Science Foundation Grant DMR0710154 and US Army Research Office under Contract W911NF0610399.

References

1. Curl, R.F., et al., *Quantum cascade lasers in chemical physics.* Chemical Physics Letters, 2010. **487**(1-3): p. 1-18.
2. Elia, A., et al., *Photoacoustic Techniques for Trace Gas Sensing Based on Semiconductor Laser Sources.* Sensors, 2009. **9**(12): p. 9616-9628.
3. Lewicki, R., et al., *Carbon dioxide and ammonia detection using 2 m diode laser based quartz-enhanced photoacoustic spectroscopy.* Applied Physics B: Lasers and Optics, 2007. **87**(1): p. 157-162.
4. Mond, M., et al., *1.9-mm and 2.0-mm laser diode pumping of Cr2+: ZnSe and Cr2+: CdMnTe.* Opt. Lett, 2002. **27**: p. 1034-1036.
5. Sorokina, I. and K. Vodopyanov, *Solid-state mid-infrared laser sources.* 2003: Springer.
6. Garbuzov, D., et al., *4 W quasi-continuous-wave output power from 2 m AlGaAsSb/InGaAsSb single-quantum-well broadened waveguide laser diodes.* Appl. Phys. Lett., 1997. **70**: p. 2931.
7. Turner, G., H. Choi, and M. Manfra, *Ultralow-threshold (50 A/cm) strained single-quantum-well GaInAsSb/AlGaAsSb lasers emitting at 2.05 m.* Appl. Phys. Lett., 1998. **72**: p. 876.
8. Chandola, A., R. Pino, and P. Dutta, *Below bandgap optical absorption in tellurium-doped GaSb.* Semicond. Sci. Technol., 2005. **20**: p. 886.
9. Rattunde, M., et al., *Comprehensive analysis of the internal losses in 2.0 mu m (AlGaIn)(AsSb) quantum-well diode lasers.* Appl. Phys. Lett., 2004. **84**(23): p. 4750-4752.
10. Haug, A., *Auger recombination in direct-gap semiconductors: band-structure effects.* Journal of Physics C: Solid State Physics, 1983. **16**: p. 4159-4172.
11. Shterengas, L., et al., *Design of high-power room-temperature continuous-wave GaSb-based type-I quantum-well lasers with lambda > 2.5 mu m.* Semicond. Sci. Technol., 2004. **19**(5): p. 655-658.
12. Hosoda, T., et al., *200 mW type I GaSb-based laser diodes operating at 3 μm: Role of waveguide width.* Appl. Phys. Lett., 2009. **94**(26).
13. Shterengas, L., et al., *Diode lasers emitting at 3 μu m with 300 mW of continuous-wave output power.* Electronics Letters, 2009. **45**(18): p. 942-U37.
14. Hosoda, T., et al., *Type-I GaSb-Based Laser Diodes Operating in 3.1- to 3.3 μm Wavelength Range.* Photonics Technology Letters, IEEE, 2010. **22**(10): p. 718-720.

WDM DEMULTIPLEXING BY USING SURFACE PLASMON POLARITONS

DJAFAR K. MYNBAEV*

Department of Electrical and Telecommunications Engineering Technology
New York City College of Technology of the City University of New York
300 Jay Street, Brooklyn, NY 11201, USA
dmynbaev@citytech.cuny.edu

VITALY SUKHARENKO

Department of Electrical and Telecommunications Engineering Technology (student)
New York City College of Technology of the City University of New York
300 Jay Street, Brooklyn, NY 11201, USA

The volume of telecommunications traffic keeps growing at an exponential rate. The optical-communications industry, the linchpin of modern telecommunications, in its quest of keeping up with this growth simply must increase the number of wavelengths in the wavelength-division multiplexing (WDM) configuration. The result of this increase would mean, too, that the number of transmitters and receivers that could be placed on one board would increase as well; hence, the density of their packaging would come to micro- and even nano-scale. At the receiver end, which we will consider in this paper, manufacturers are now able to fabricate an array of photodiodes (PDs) on a single wafer, reducing the size of an individual PD to hundreds of nm. The main obstacle in approaching this scale from a communications link point of view, however, is the diffraction limit. One possible solution to this problem is the use of plasmonics. This paper discusses a possible approach to using surface plasmon polaritons (SPPs) for WDM demultiplexing, presents a possible scheme to implement this approach, and offers our analysis of this scheme along with suggestions for future developments.

Keywords: Surface plasmon polaritons; WDM demultiplexing.

1. Introduction

Surface plasmon polaritons (SPPs) are two-dimensional waves that propagate between conductors (metals) and dielectrics. These surface waves are excited when light strikes the dielectric-metal interface; the energy of the photons is transferred to the metal and resonantly excites the oscillations of free electrons. The electrons' response results in the creation of dynamic charges on the metal's surface; these charges, in turn, produce waves called SPPs [1]. Fig. 1 illustrates these explanations. SPPs can be considered, in essence, secondary EM radiation obtained in response to incident light.

* New York City College of Technology, 300 Jay Street, Brooklyn, NY 11201, USA.

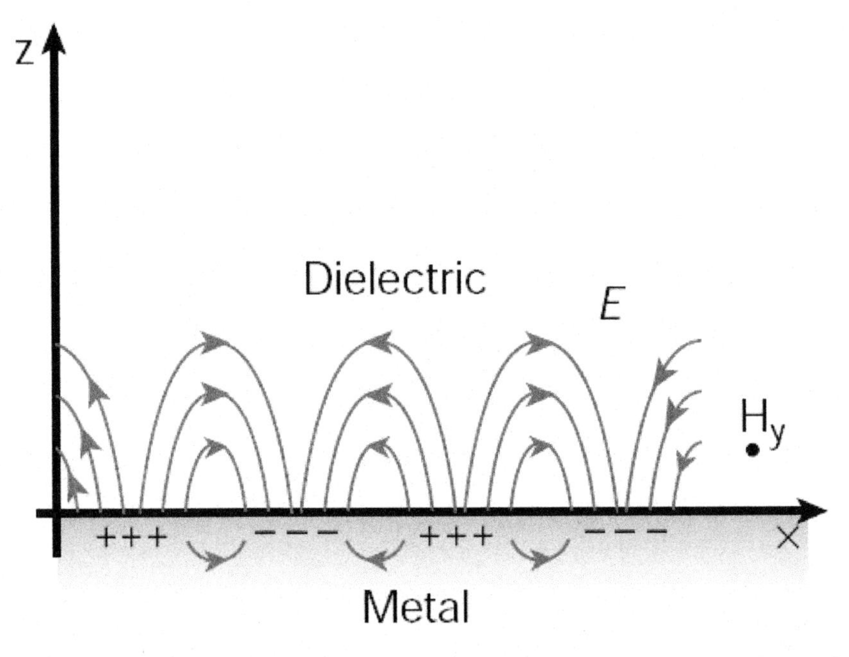

Fig. 1. This Surface plasmon polaritons: Electron charges and excited electric waves. (Reprinted with permission of Igor Zozulenko[2].)

The term *surface* refers to the waves that travel along the metal surface (between the metal and a dielectric) in contrast to waves excited in the metal volume. The term *plasmon* stems from the word *plasma*, a charged gas describing the behavior of free electrons according to the Drude model of metals. *Polariton* is a quasiparticle—half wave and half matter—well defined in physics.

SPPs have been well known for more than a century; today's strong interest in the study of this phenomenon has been due to the latest advances in nanotechnology, advances that now make applications of SPPs in numerous fields quite feasible. The main theory behind the excitation of SPPs, as well as the techniques for their excitation, is well known[1]; we will consider here only the excitation of SPPs with diffraction gratings.

2. The Frequency of SPPs

We consider here a thin metal film placed between two dielectrics with corrugations on both sides of the film; thus, we have top and bottom metallic gratings. In this paper, we consider light falling on the top grating and SPP radiation emanating from the bottom grating. Fig. 2 shows an example of such an arrangement and excited SPPs[3]. When the metal strip is thicker than 100 nm, only a single SPP mode gets excited; when, however, the metal film is thinner than 100 nm, two SPP modes—short-range and long-range—are excited.

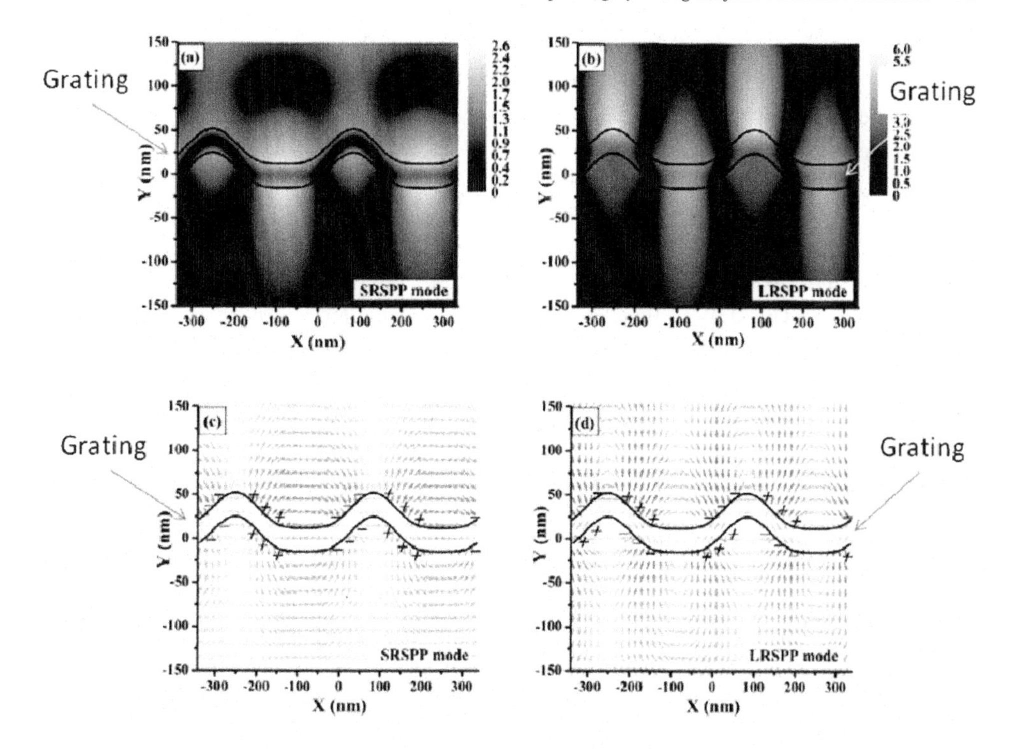

Fig. 2. SPP modes excited on a thin metal strip corrugated on both sides (metal gratings): (a) and (b) show the actual pictures of the excited SPPs; (c) and (d) schematically show charges and excited SPP modes. SRSPP and LRSPP stand for short-range SPP and long-range SPP, respectively. (Reprinted with permission from[3].)

Thus, in general, a thin metal strip with gratings might acts as a transducer of incident photons to SPPs. The goal of this section of our paper is to present the frequency of the SPPs and its dependence on parameters of the materials and the gratings.

In response to the incident light, the free electrons that form the plasma oscillate at their own natural *plasma frequency*; this frequency lies in the ultraviolet spectrum. The plasma's frequency, ω_p, can be calculated from the following equation[1]

$$\omega_p = n\, e^2/\varepsilon_0\, m, \tag{1}$$

where n is the density of a free electron gas that oscillates with respect to fixed positive ions, m is the effective optical mass, e is the charge of the electron, and ε_0 is the permittivity of vacuum. It's worth noting that the oscillation of the free electrons is dumping due to collisions; the characteristic collision frequency, γ, is equal to 100 THz at room temperature. The dielectric function of a metal, $\varepsilon(\omega)$, is given by[1]

$$\varepsilon(\omega) = 1 - \omega_p^2/(\omega^2 + i\gamma\omega), \tag{2}$$

where ω is the frequency of the incident radiation.

The key to the investigation of the excitation of the SPPs is the relationship between their propagation constant, β, and the frequency, ω, of the incident light. At the interface between the metal of infinite optical thickness and a dielectric, the propagation constant can be represented as[1]

$$\omega/c^2 \sqrt{\varepsilon(\omega)\,\varepsilon_2/(\varepsilon(\omega)+\varepsilon_2)} = \beta, \tag{3}$$

where ε_2 is the electric permittivity of a dielectric. Fig. 3 visualizes the relationship between β and ω. The straight lines represent lines of light. Therefore, in order to couple the SPPs, β has to be greater than the propagation constant of light in the given dielectric. This necessary condition for exciting SPPs leads to the need to use grating or prism coupling. We consider gratings because they are more suitable to our application.

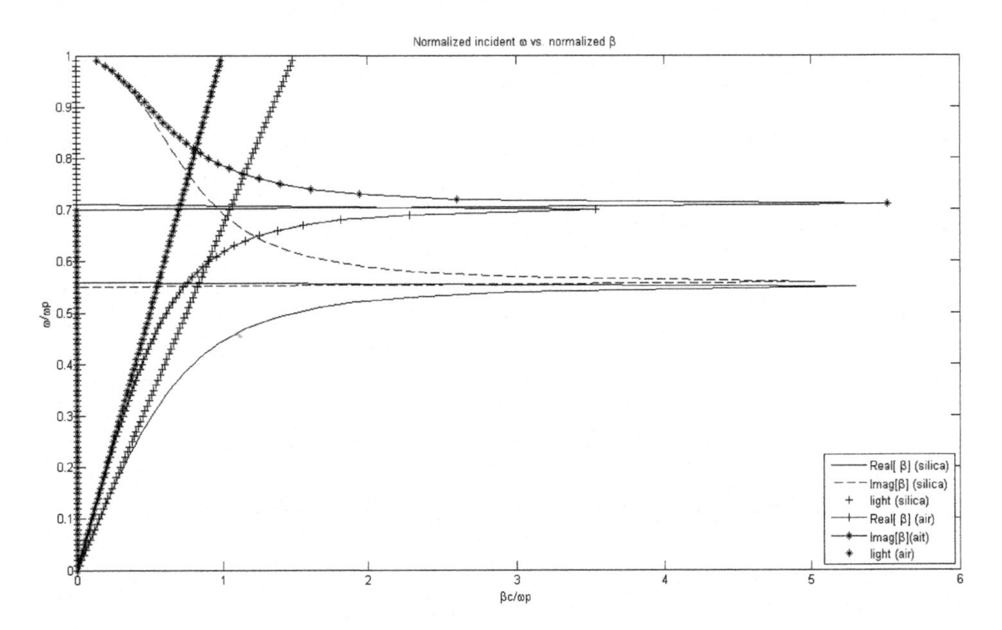

Fig. 3. Normalized ω vs. normalized β, as described by Eq. 3. (Here $\varepsilon_{silica} = 2.25$ and $\varepsilon_{air} = 1$.)

It's worth noting that, in general, at the large value of propagation vectors SPPs frequency approaches characteristic surface plasmon frequency[1]

$$\omega_{sp} = \omega p/\sqrt{(1 + \varepsilon_{dielectric})}. \tag{4}$$

In our arrangement, the plasmonic frequency, ω_{sp}, strongly depends on the incident frequency and the top grating — the confinement of the thin corrugated metal strip and the top dielectric. Similarly, the radiated frequency strongly relates to the plasmonic frequency and the bottom grating. These relationships can be seen from the following equations, where Eq. 5 governs the excitation[1] and Eq. 6 — radiation[4]:

$$n\,\omega/c^2 \sin\theta \pm vg = \beta_{top} \tag{5}$$

and

$$\pm n\,\omega/c^2 \sin\varphi = \pm vg \pm \beta_{bottom}, \tag{6}$$

where v is the order number, $g = 2\,\pi\,/\,grating\ pitch$ is the grating vector, n is the refractive index of the top (incident) dielectric, θ is the angle of incidence, φ is the angle of radiation, and β is the propagation constant of SPPs. Eq. 5 and Eq. 6 show the relationship among the propagation constant, the frequency of the incident light, the radiated frequency, and pitches of the top and bottom gratings.

3. The Concept of WDM Demultiplexing with SPPs

We have investigated the concept of WDM demultiplexing by using SPPs: The transmission optical fiber is shaped into a cone to increase light intensity at the tip. The corrugated thin silver film is placed perpendicular to the tip and submerged in dielectric material on both sides with a different electric permittivity on each side. Incident light is launched onto the entire surface of the top grating; thus, SPP excitation occurs almost uniformly across the entire metal film. The bottom grating, from which SPP radiation is collected, has variable pitches; thus, the relationship between the excited SPP frequency and the radiated frequency changes across the metal film. This scheme is the subject of our discussion today. Here's how we conducted our study:

First, we investigated how the radiation frequency follows the change of incident ω and the bottom-grating pitch. In these calculations we assumed the top (from the top dielectric to the metal) incident angle is constant and equal to 22.2^0. Then we computed the angle of transmission within the metal by means of Snell's law. We also used Snell's law to compute the angle at which electromagnetic power radiated from the first-order bottom grating into the bottom dielectric. Another assumption was that our metal is silver whose permittivity is given by[4]

$$\begin{aligned}\varepsilon = &-225.3189 + 1.9863*10^{-13}\omega - 6.0794*10^{-29}\omega^2 + 8.3810*10^{-45}\omega^3 - 4.3004*10^{-61}\omega^4 \\ &+ i(832575 - 1.3279*10^{-13}\omega + 9.0474*10^{-29}\omega^2 - 3.2880*10^{-44}\omega^3 + 6.659*10^{-60}\omega^4 \\ &- 7.0893*10^{-76}\omega^5)\end{aligned} \tag{7}$$

Yet the other assumption was that $\Delta\beta$ is considered a difference between β_0, given in Eq. 3 (a propagation constant at the confinement between the top dielectric and the optically infinite metal), and β, defined by Eq. 5 (which represents propagation constant matching at the confinement between the top dielectric and metal). The more accurate value of $\Delta\beta$ has to incorporate the thickness of the metal strip, thus enabling us to calculate LRSPP and SRSPP at specific condition. However, if the thickness of the metal strip is 100 nm or greater, LRSPP and SRSPP have almost identical frequencies[1]. Thus, we assumed our metal film to be about 100 nm in thickness. Fortunately, the main idea of the proposed WDM demultiplexing scheme does not depend heavily on this assumption. Under these assumptions and determining the top grating pitch equal to 380 nm, we were able to construct Fig. 4, showing the reflection coefficient as a function of the incident frequency. In Fig. 4, Graph 1 has been built by using Eq. 9b, whereas Graph 2 has visualized Eq. 9a.

$$r_p = n_2 \cos \varphi_t - n_1 \cos\varphi\beta / n_2 \cos \varphi_t + n_1 \cos\varphi\beta, \tag{9a}$$

where φ_t - angle of transmittance, $\varphi\beta$ - angle of incidence, n_1 - top (incident) medium refractive index, n_2 - bottom (transmitted) medium refractive index.

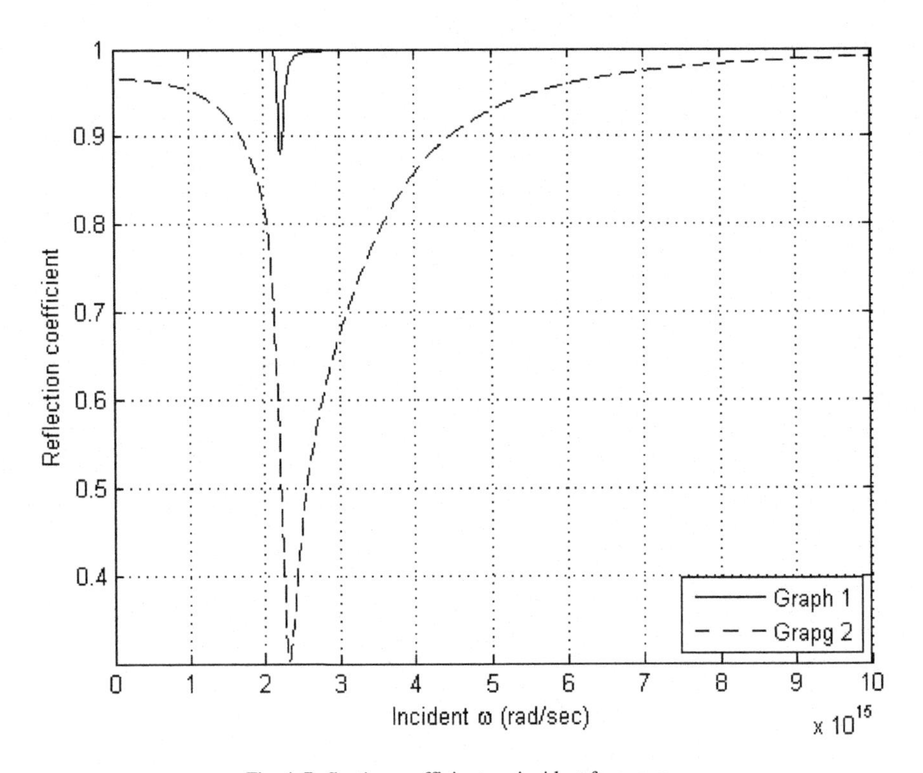

Fig. 4. Reflection coefficient vs. incident frequency.

$$R = 1 - (4\, Im[\beta_0]\, Im[\Delta\beta]) / (\beta - (\beta_0 + \Delta\beta))^2 + (Im[\beta_0] + Im[\Delta\beta])^2 \tag{9b}$$

One can see that at $\omega \approx 2.3$ rad/s light penetration into the top metal grating became much more pronounced, which resulted in more effective SPP excitation at this frequency. It's quite interesting to observe that the both graphs show a drop at almost the same frequencies.

The propagation constant, β, given in Eq. 9b, must have excitation at frequencies where the graph drops below the 0.99 coefficient of reflection; therefore, only these frequencies will excite SPPs. At these frequencies, radiation at the bottom grating must take effect; thus, altering the grating pitch of a bottom grating will result in altering the radiated frequency. Different incident frequencies will correspond to different β's. Imagine the entire surface of the bottom silver grating divided into sections with different pitches. Fig. 5 visualizes this effect showing a grating with two different pitches.

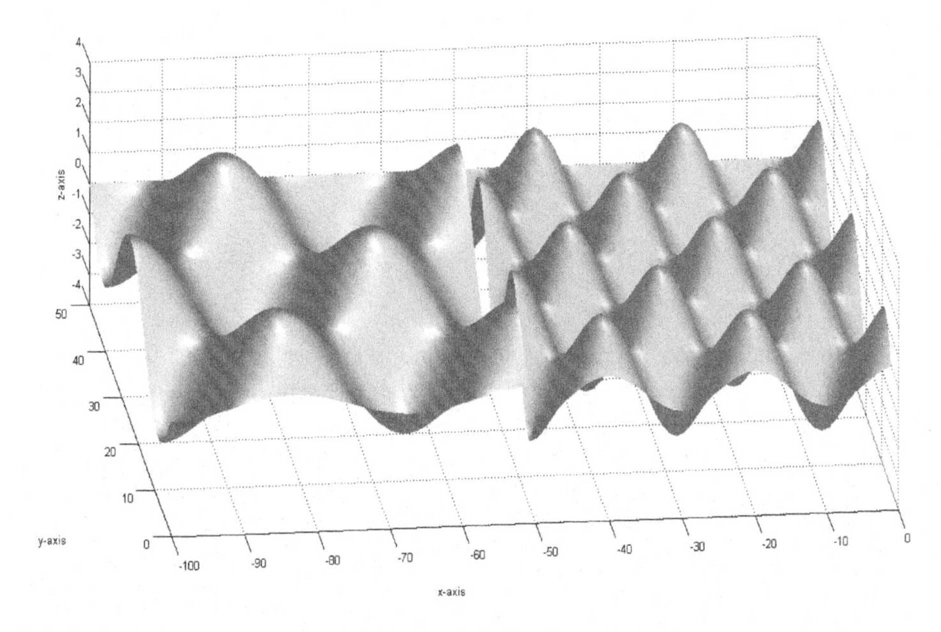

Fig. 5. A grating surface with different pitches.

If the first section of the grating has, for example, a pitch of 360 nm, then at an incident frequency of $2.316*10^{15}$ rad/sec with specific β, the radiated frequency will be equal to $9.8*10^{15}$ rad/sec. At the same time, if the second section has a pitch of 380 nm, then at a different incident frequency of $2.275*10^{15}$rad/sec, the radiation frequency will be equal to $9.8*10^{15}$ rad/sec, exactly the same frequency as in the first section with pitch 360 nm. Fig. 6 whose graphs were built based on Eqs. 5, 6 and 9b explicitly demonstrates this phenomenon.

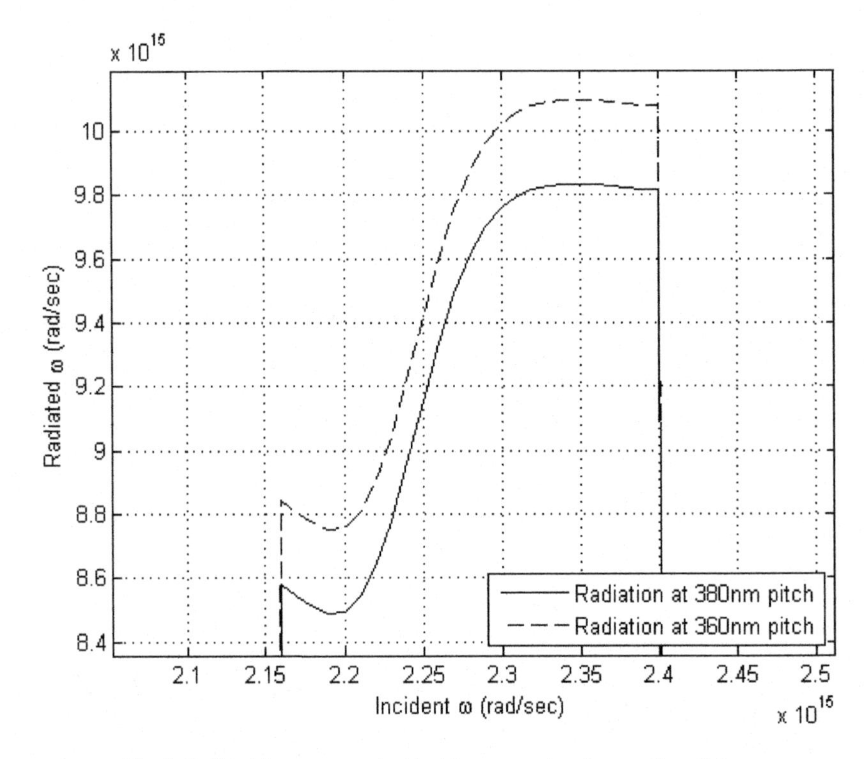

Fig. 6. Radiated frequency vs. incident frequency at various grating pitches.

Therefore, altering the pitches of the bottom grating, we can construct the metal strip, which will radiate exactly the same frequency at different locations on the bottom surface in response to different incident frequencies. Simply knowing from which section plasmonic radiation is received, we can uniquely identify the incident frequency. Practically speaking, we needed to place PDs at different points of the bottom grating and identify their positions with respect to the specific pitches. We also needed to place an optical filter between the bottom grating and the PD array that allows for the transmission only one radiated frequency, 9.8*1015 rad/s in our example. Then, identifying from which PD the signal is collected, one can find out from which incident wavelength this signal originates. Thus, this scheme works as a WDM demultiplexer at the nanometer scale. Fig. 7 illustrates the construction of this scheme.

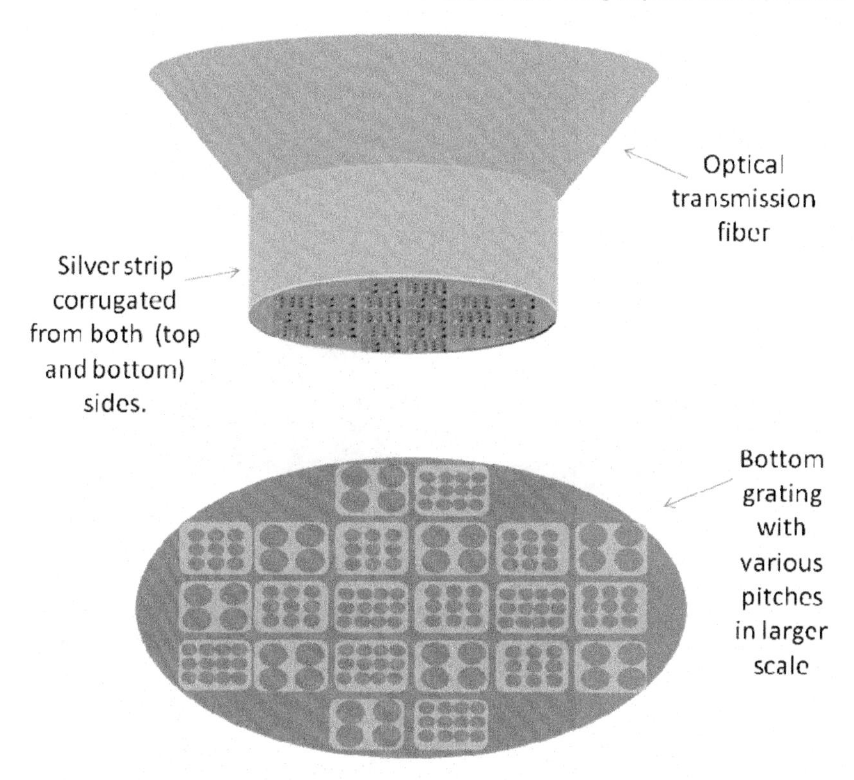

Fig. 7. WDM demultiplexer arranged at the tip of the coned optical transmission fiber.

The proposed scheme puts all PDs in a very comfortable regime because they constantly work at the same frequency; it can be easily arranged for this frequency to be at the maximum sensitivity of the PDs.

Fig. 8 offers a demonstration of the dependence of radiated frequency on both the incident frequency and the bottom pitch.

Radiated frequency also depends on the top-grating pitch, as shown in Fig. 9. At different incident pitches, the SPPs β's at the confinement between the silica and silver will change too. This could be a quite useful phenomenon, which would allow for greater difference between the radiated frequencies, but it's a subject for further study.

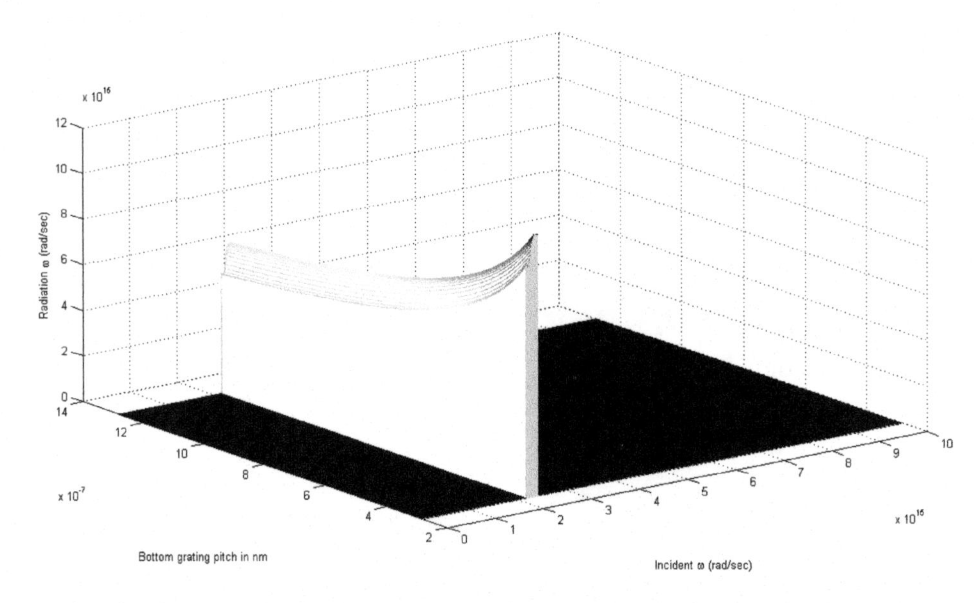

Fig. 8. Radiated frequency as a function of both the incident frequency and the bottom pitch at 0.99 reflection coefficient.

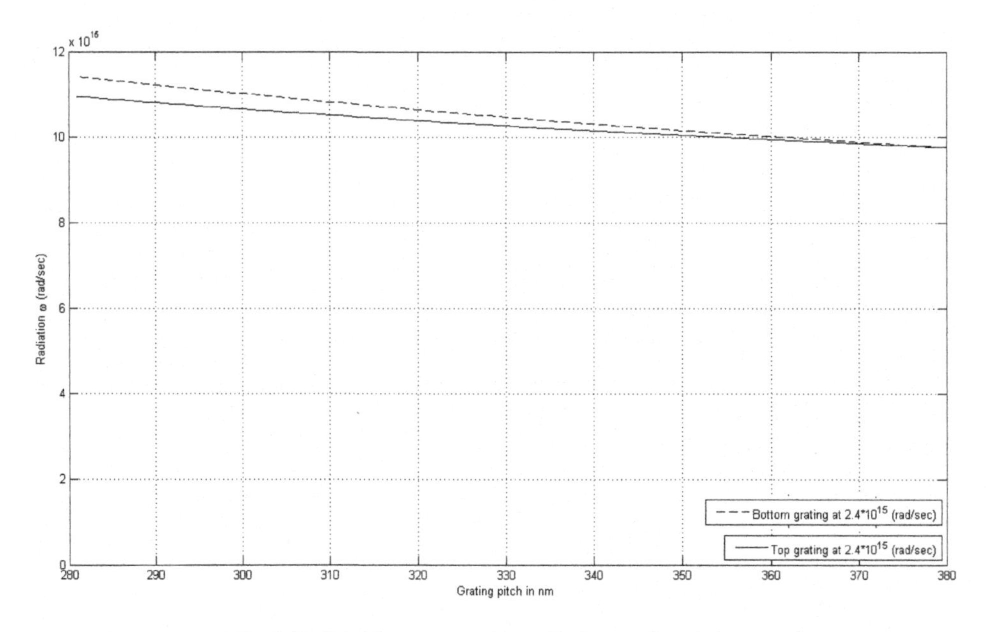

Fig. 9. Radiated frequency vs. top and bottom grating pitches.

4. Summary

We propose an optical scheme that can serve as a WDM demultiplexer in nano scale. Calculations show that realization of this scheme is quite possible. However, much work has to be done—both theoretically and experimentally—to develop this scheme from a practical standpoint.

Acknowledgment

This work was supported by PSC-CUNY grant 61558-00 39 and by the Emerging Scholars program of the New York City College of Technology.

References

1. Stefan A. Maier, *Plasmonics: Fundamentals and Applications*, (Springer Science+Business Media LLC, 2007).
2. Igor Zozoulenko, Surface plasmons and their applications in electro-optical devices, Solid State Electronics, Department of Science and Technology, Linköping University, Sweden, (2006). http://www.itn.liu.se/meso-phot
3. Z. Chen, I.R. Hooper, and J.R. Sambles, Coupled surface plasmons on thin silver gratings, *Journal of Optics A: Pure and Applied Optics*, **10** (2008), 015007.
4. S. Wedge, I.R. Hooper, I. Sage, and W.L. Barnes, Light emission through a corrugated metal film: The role of cross-coupled surface plasmon polaritons, *Physical Review B* **69**, 245418 (2004).
5. I.R. Hooper and J.R. Sambles, "Surface plasmon polaritons on thin-slab metal gratings," *Physical Review B* **67**, 235404 (2003).

SILICON AND GERMANIUM ON INSULATOR AND ADVANCED CMOS AND MOSHFETs

CONNECTING ELECTRICAL AND STRUCTURAL DIELECTRIC CHARACTERISTICS

G. BERSUKER, D. VEKSLER, C. D. YOUNG, H. PARK, W. TAYLOR, P. KIRSCH and R. JAMMY

SEMATECH, 2706 Montopolis Dr.,
Austin, TX 78741, USA
gennadi.bersuker@sematech.org

L. MORASSI, A. PADOVANI and L. LARCHER

DISMI Università di Modena e Reggio Emilia and IU.NET, 42100 Reggio Emilia,
Modena, Italy

An attempt is made to correlate electrical measurement results to specific defects in the dielectric stacks of high-k/metal gate devices. Defect characteristics extracted from electrical data were compared to those obtained by ab initio calculations of the dielectric structures. It is demonstrated that oxygen vacancies in a variety of charge states and configurations in the interfacial SiO_2 layer of the high-k gate stacks contribute to random telegraph noise signal, time-dependent dielectric breakdown, and the flatband voltage roll-off phenomenon.

Keywords: random telegraph noise; dielectric breakdown; flatband roll-off.

1. Introduction

Relentless device scaling challenges the traditional "empirical" approach to device characterization: increasing variability adversely affects the predictive capability of well-established statistical evaluation methods, while new materials and complex multi-component gate stack structures may result in instabilities from process-generated defects (rather than stress-generated ones, which are usually associated with the time dependency of device characteristics). These factors point to a growing need to identify the nature of defects affecting the electrical characteristics of devices, specifically reliability, that would allow developing physics-based degradation models, as well as provide helpful feedback to process optimization efforts.

In this study, we correlate electrical characteristics of metal/high-k gate stacks that contain a SiO_2 layer at the gate dielectric/substrate interface to specific atomic defects in this interfacial layer. As has been demonstrated by electrical, physical, and modeling studies, Hf-based high-k films can modify the stoichiometry of the underlying SiO_2 layer by rendering it oxygen-deficient. This leads to an increase in its dielectric constant and a higher density of fixed charges in this layer, which degrade channel carrier mobility. Here we focus on identifying specific characteristics of the oxygen vacancy defects by matching defect structural parameters extracted from electrical measurements to those obtained by ab initio calculations.

By using the analysis for low frequency noise data, which takes into consideration multi-phonon relaxation processes induced by the charge trapping/detrapping in the dielectric, we demonstrate that the essential characteristics of traps in the SiO_2 layer in high-k devices can be obtained. Strong dependency of the electron capture/emission times on defect relaxation energy allows extracting the latter value, which can be used as a defect identifier along with the defect energy and capture cross-section characteristics. Complementary modeling of the gate leakage current in high-k devices during electrical stress using the same approach yields characteristics of the traps in the interfacial SiO_2 layer contributing to the trap-assisted tunneling process (TAT). Comparing defect energy characteristics from random telegraph signal noise (RTN) and TAT measurements to those obtained by *ab initio* calculations the electrically active defects can be tentatively assigned to oxygen vacancies. Based on these findings, as well as an earlier transmission electron microscopy/electron energy-loss spectroscopy (TEM/EELS) study of the elemental composition of the breakdown path, we propose that the breakdown path formation/evolution in the interfacial layer is associated with the growth of an oxygen-deficient filament facilitated by the grain boundaries of the overlying high-k film.

Interfacial layer defects were also found to control the so-called flatband voltage (V_{fb}) roll-off (R-O) phenomenon, which significantly limits the available options for metal/high-k transistor fabrication. This phenomenon describes a significant reduction of the effective work function (EWF) values of gate stacks consisting of a metal electrode, high-k dielectric, and interfacial SiO_2 layer when the thickness of the latter layer is scaled down. It was determined that the oxygen vacancies at the interface with the Si substrate, generation of which occurs more effectively during high temperature processing and in thinner interfacial SiO_2 layers, may acquire positive charge resulting in lower V_{fb}.

In the subsequent sections, we discuss approaches to identify the interlayer (IL) defects contributing to RTN, time-dependent dielectric breakdown (TDDB), and flatband voltage roll-off.

2. Noise Generating Defects in High-k Gate Stacks

2.1. *Analysis of Random Telegraph Signal Noise (RTN)*

Electrically-active defects in the gate dielectric are capable of generating noise (small fluctuations in the device output current [Fig. 1]). The phenomenon caused by a carrier in the channel hopping in and out of a single defect and resulting in the output current fluctuation between two discrete values, is called random telegraph signal noise (RTN) [1] (charge exchange with more than one trap results in a multilevel RTN). In large area devices containing sufficiently high density of such traps uniformly distributed through the dielectric thickness, overlapping signals from the individual traps form the 1/f dependency of the noise power spectral density.

RTN provides information about the trap average capture and emission times, Fig. 1. Conventional RTN analysis is based on the ratio of these times and employs a detailed balance principle [2,3,4]. This analysis allows such characteristics as the trap thermal ionization energy and spatial position to be extracted while avoiding explicitly considering an electron trapping/detrapping mechanism. However, knowing the defect energy alone is not sufficient for elucidating the nature of the defect. To obtain additional defect characteristics, the analysis should be based on a wider set of independent experimental parameters contained in measured RTN data, specifically, the data sets of both trap capture times and emission times and their dependences on the electrical biases and temperature. For this purpose, we must describe the physical processes of electron trapping and detrapping.

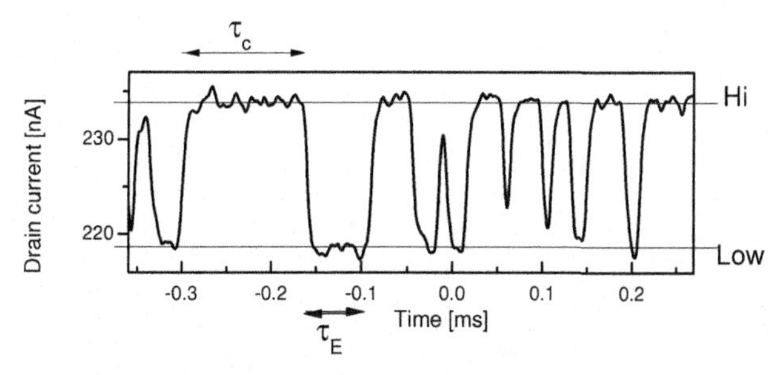

Fig. 1. Typical two-level RTS signal in a high-k MOSFET. The low current state corresponds to the trap being filled; the high current state corresponds to the empty trap. The time periods in the low, τ_E, and high, τ_C, current states correspond to the electron emission and capture times, respectively.

Electron transitions in condensed media using the concept of multiphonon-assisted (non-elastic) processes (see, for instance, [5,6,7] and references therein) were considered for a variety of physical phenomena. The specific case of electron trapping at the bulk defect in the gate dielectric can be considered, within this approach, as a combination of two coherent processes [8]:

1) Electron tunneling from the transistor channel to the trap (Fig. 2a), with the probability

$$P_{tun} = \exp(-x_T / \lambda_e) \tag{1}$$

2) Rearrangement of the lattice atoms forming the trap (Fig. 2b) to accommodate an additional electron charge, with the probability [6,7]:

$$P_{relax} = e^{-(2\bar{n}+1)S} \sqrt{\left(\frac{\bar{n}+1}{\bar{n}}\right)^p} I_p\left(2S\sqrt{\bar{n}(\bar{n}+1)}\right) \tag{2}$$

Here x_T is the trap distance from the substrate; λ is the characteristic electron tunneling length; $I_p(x)$ is a Bessel function of an order p; $\bar{n}$ is an equilibrium number of phonons; $p = E_0/\hbar\omega$, with E_0 being the total energy difference between the initial (trap is empty) and the final (electron in the relaxed trap) states of the system, $S = E_{relax}/\hbar\omega$ is the Huang-Rhys factor, and E_{relax} the energy associated with the displacements of the atoms in the dielectric (the trap relaxation energy) caused by electron trapping; ω is the characteristic phonon frequency associated with these displacements; and T is an ambient temperature.

The second process, called the "structural relaxation," might significantly affect the trapping dynamics. Indeed, to form a new atomic configuration around the trapped electron, the lattice atoms must shift from their initial equilibrium positions, which is, in general, associated with the system overcoming an energy barrier [5,6,8]:

$$E_B=(E_{relax}-E_0)^2/4\ E_{relax}\ kT \tag{3}$$

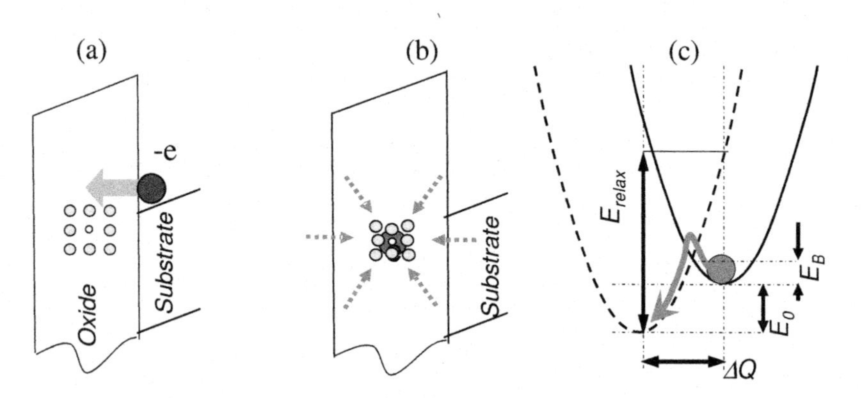

Fig. 2. Schematic of the band diagram illustrating (a) electron tunneling to the trap and (b) trap relaxation caused by the electron trapping. (c) Total energy diagram, taking into account electon and phonon subsystems. Q – the generalized coordinate of the atomic displacements in the system. Solid and dashed parabolas correspond to the full system potential energy in the empty and filled traps, respectively. E_{relax} is the energy corresponding to displacements of the lattice atoms ΔQ. E_0 is the energy loss by an electronic subsystem after a trapping event. The diagram indicates that electron trapping is associated with the system transitioning over the barrier E_B.

This barrier is responsible for retarding the electron trapping (and detrapping) process (Fig. 2c). In the case of $S >> p$ and $kT >> \hbar\omega$,

$$P_{relax} \approx e^{-E_B/kT} \tag{4}$$

Within the Shockley Read Hall (SRH) approximation, the capture and emission times can be expressed as follows:

$$\tau_c = (v_t \times n \times \sigma_0 P_{tun} P_{relax})^{-1},$$
$$\tau_E = \tau_c \exp\left(-((E_T + e \cdot F_{ox} \cdot x_T) - E_F)/k \cdot T\right), \tag{5}$$

Here v_t is an electron thermal velocity, n is electron concentration in the substrate, σ_0 is the electronic component of the trap capture cross-section, E_F is the Fermi energy in the substrate, F_{ox} is the electric field in the oxide, e is the electron charge, and E_T is the trap energy.

A comprehensive analysis, which considers the total system energy including phonons in the dielectric along with the electronic subsystem, allows the measured values for the capture/emission times to be connected to the trap's relaxation energy. The latter represents an important trap characteristic serving as a marker of the defect's structure, in addition to the trap energy, location, and capture cross-section.

2.2. *Extracting defect characteristics*

For high-k gate stacks, the noise data were collected on small (W/L = 0.3 μm/0.1 μm) transistors with a 1 nm SiO_2/3 nm HfO_2/TiN gate stack, in the temperature range of 300K–345K, and capture/emission times were extracted. The dependence of the times versus the electrical bias and temperature were reproduced theoretically using Eqs. (1-5), see Fig. 3. The traps in the high-k MOSFETs under investigation were found to reside in the interfacial SiO_2 layer, sandwiched between the HfO_2 and Si substrate, around 0.3 nm from the substrate interface; the trap energy was ~ 3 eV, counting from the conduction band edge of the SiO_2 dielectric. At such short tunneling distances, the capture and emission times are controlled by the structural relaxation process (Eq. (2)) with the relaxation energy of ~ 1.7–1.9 eV.

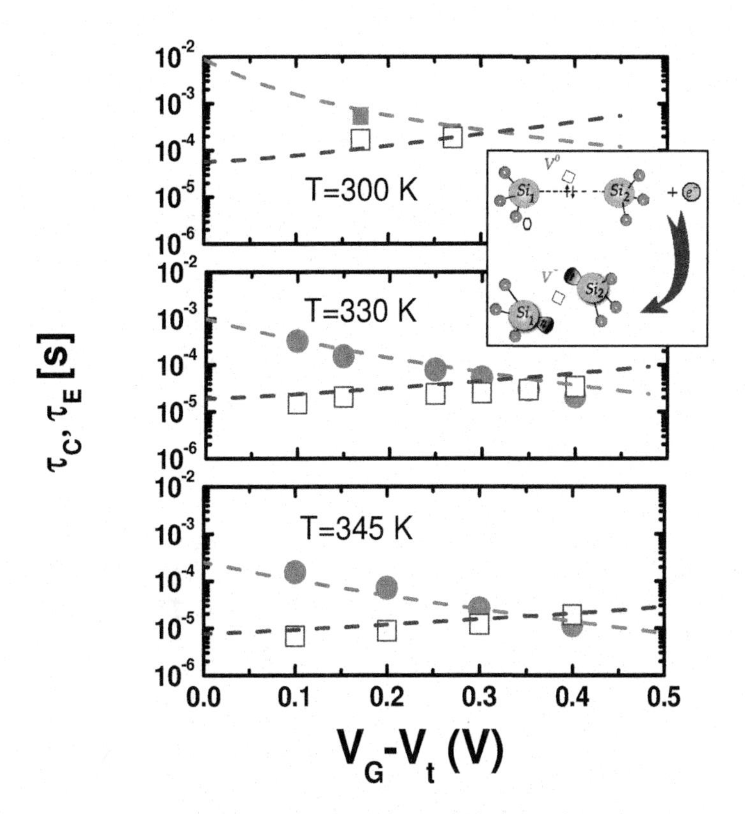

Fig. 3. Experimental (dots) and calculated (lines) average capture (circles) and emission (squares) times for a 1 nm SiO_2/3nm HfO_2/TiN nFET. T = 300 K, 330 K, and 345 K. A theoretical fit is performed with a single set of trap parameters for the entire range of gate biases and temperatures. The inset shows schematically an electron capture-induced conformation of a neutral oxygen vacancy [9].

The extracted defect parameters, including the trap relaxation energy and the measure of the trap lattice distortion upon electron capture, were compared to the results of *ab initio* calculations of several possible configurations of oxygen vacancies in SiO_2 [9]. The comparison allows the traps contributing to noise to be tentatively identified as neutral (V^0) oxygen vacancies (converted into V^- after electron trapping) in the interfacial SiO_2 layer of the MOSFET gate stack. V^- vacancy defects with similar characteristics were also identified from the RTS data collected on transistors with SiON/poly-Si gate stacks [10].

Note that the practical available frequency range of the noise measurement setup limits the observable spatial and relaxation energy window: the traps with lower E_{relax} can be observed farther away from the interface with the Si substrate (to keep the characteristic times above the low limit of the available measurement frequencies) [10].

3. Defects Responsible for High-k Gate Stack Degradation/Breakdown

Previous studies demonstrated a strong correlation between defect generation in the interfacial SiO_2 layer (IL) in high-k stacks under stress (as measured by the frequency-dependent charge pumping method, f-CP) [11,12] and an increase of the gate leakage current measured at low gate voltages $V_g \ll V_{stress}$ (a so-called stress-induced leakage current [SILC]). On the other hand, SILC evolution during stress correlates to the gate stack transition through the various degradation stages, soft and progressive breakdown (BD) phases [13,14] suggesting that the IL degradation might be primarily responsible for the overall degradation of the high-k gate stack. Below we, first, show evidence that the IL in the fully processed transistor gate stack is highly oxygen deficient and then discuss the formation/evolution of the BD path in the IL as associated with the stress-induced growth of an oxygen-deficient filament facilitated by pre-existing (as-processed) O-vacancies.

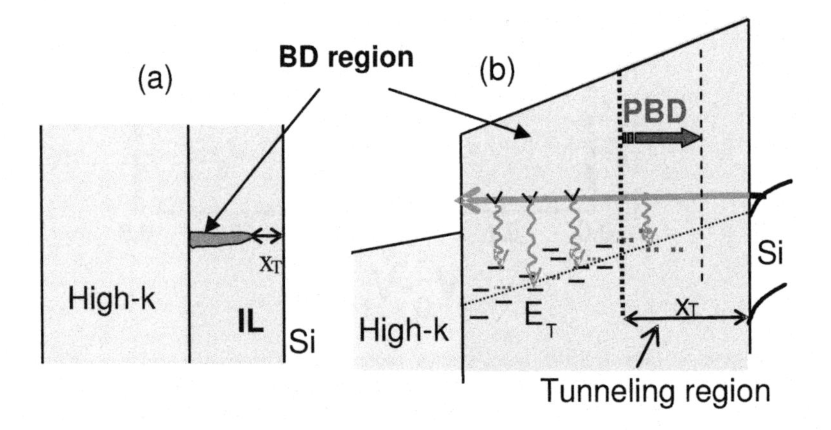

Fig. 4. Schematic illustration of the proposed IL BD model. (a) Formation of the O-deficient filament in IL in the result of SBD. (b) Band energy diagram of the SBD region. PBD is associated with propagation of the BD filament toward Si substrate accompanied by generation of additional O-vacancy states (dotted lines). The wavy line indicates trap relaxation caused by an electron capture. x_T is the minimum electron tunneling distance through the unbroken IL portion, E_T is the energy of the relaxed traps.

3.1. *Physical evidences of O-vacancies in IL*

An increase in the density of oxygen vacancies in the interfacial 2 nm thermal SiO_2 layer in the 3 nm atomic layer deposition (ALD) HfO_2 gate stack after high temperature annealing (1000°C/10 sec in N_2 ambient) was detected using high angle annular dark field scanning transmission electron microscopy (HAADF-STEM) in combination with EELS [15]. Si $L_{2,3}$ edge signals collected in the SiO_2 region show the signature of the Si atoms in various coordination states while only a Si^{4+} signal, as expected for the stoichiometric SiO_2, was observed in the as-deposited sample. This result was confirmed by Si 1*s* variable kinetic energy X-ray photoelectron spectroscopy (VKE-XPS) data, while no Hf silicate formation was detected by Hf 4*f* spectra at different electron take-off angles that provided an Hf distribution profile [16].

Additionally, samples with a 3 nm ALD HfO_2/TiN stack deposited on a 1 nm SiO_2 layer exhibit a dramatically higher density of the E' centers after the 1000°C anneal compared with the as-deposited high-k stack [17], which is consistent with the process-induced mechanism of the oxygen vacancy generation in SiO_2 films in the high-k gate stack. Annealing the stack comprised of the oxygen-deficient ALD HfO_2 film (deposited using shorter oxidation cycles) and 1.1 nm SiO_2 interfacial film resulted in an order of magnitude increase in the density of the oxygen vacancies in the SiO_2 layer. This demonstrated that higher density of oxygen vacancies in the high-k dielectric induces more oxygen vacancies in the underlying SiO_2 layer [15].

Recent TEM/EELS results [18] reveal that soft breakdown (SBD) in a thin SiO_2 gate dielectric is associated with the formation of a highly oxygen-deficient filament. One may expect that the presence of an oxygen-deficient region would facilitate the formation of a BD path in the near the high-k film

Based on the physical characterization data presented here, we propose a model that assumes that a highly oxygen-deficient filament formed during a SBD event (as in SiO_2 [18]) consumes only a portion of the IL oxide adjacent to the high-k, Fig. 4. Within this BD region, the band gap is effectively collapsed due to a high density of localized states associated with unpaired Si 3p electrons (since oxygen is lacking, similar to the well-studied oxygen-deficient transitional SiO_2 layer near the SiO_2/Si interface [19]). These unoccupied states help transfer the electrons through the dielectric.

3.2. *Defects identification*

To verify the proposed BD mechanism and identify defects in the SiO_2 IL responsible for the observed SILC variations (which are presumed to reflect on IL degradation) temperature-dependent current-voltage (I-V) characteristics were collected on fresh devices and during stress after the SBD, progressive BD (PBD) and hard BD (HBD) phases, Fig. 5. The corresponding activation energy values were extracted and are plotted in Fig. 6. The activation energies exhibit a peculiar characteristic of first sharply increasing after SBD and then gradually decreasing with further dielectric degradation.

Based on the above breakdown model, gate leakage-gate voltage (I_g-V_g) data were simulated using the electron transport model, which considers a multi-phonon trap-assisted tunneling mechanism (as discussed in 2.1), including random defect generation and barrier deformation induced by the charged traps [20], Fig. 7.

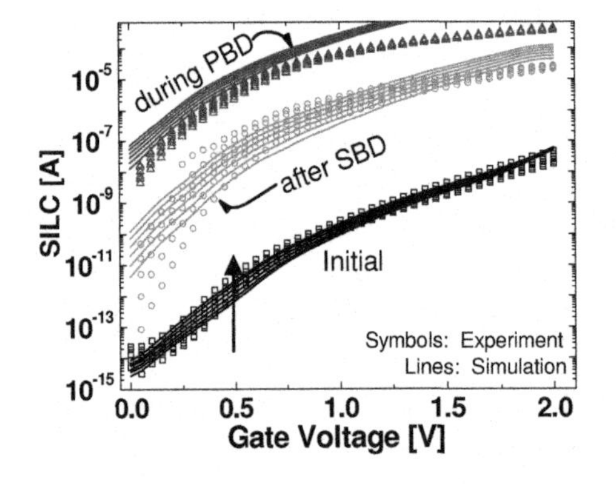

Fig. 5. Temperature dependency of SILC measured before stress, after SBD, and during PBD in 1.1 nm SiO_2/3 nm HfO_2/TiN transistors. Each group of symbols represents a set of I-V data measured within 25°C-150°C. Lines are simulation results.

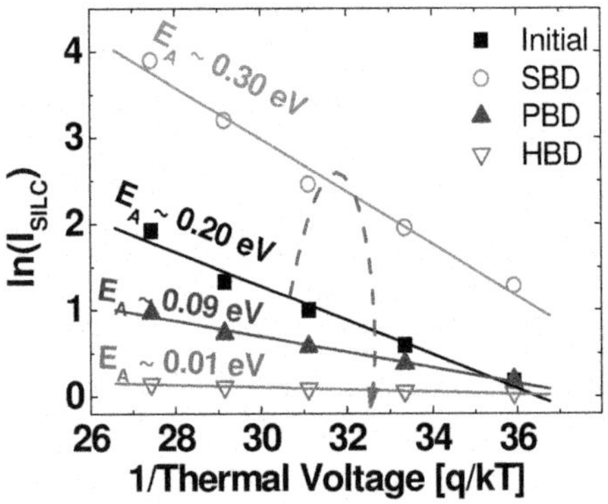

Fig. 6. Activation energies of SILC measured before stress, after SBD, during PBD, and after HBD in 1.1 nm SiO_2/3 nm HfO_2/TiN transistors.

Simulations quantitatively describe both SILC temperature dependency and exponential current increase at different stages of BD, Fig. 5, as well as the corresponding SILC activation energies, Fig. 8. The energy characteristics of the trapping centers extracted by fitting the electron transport equations to the temperature-dependent leakage current data in Fig. 6 are presented in Table I (note that the E_b values were calculated using Eq. (3)).

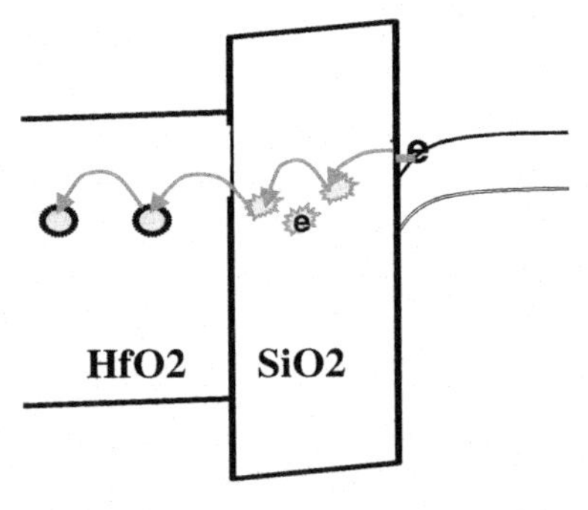

Fig. 7. Schematic of the leakage processes through the dielectric stack considered by the SILC simulations.

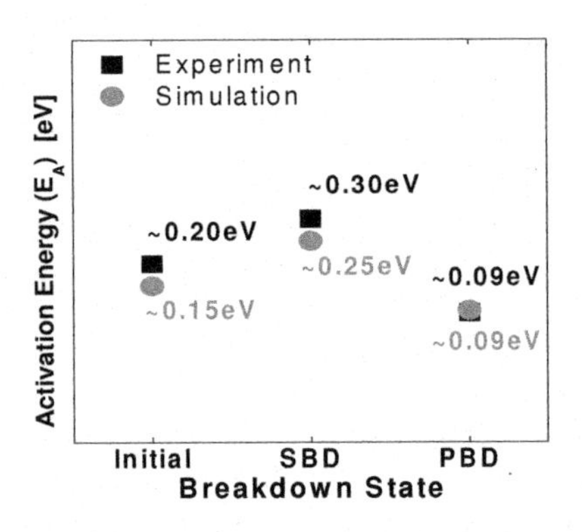

Fig. 8. Measured (Fig. 6) and calculated (Eq. 3) SILC activation energies at different stages of the stress-induced gate stack degradation.

According to the simulations, the stress-induced increase of the TAT current and its temperature dependency before SBD can be assigned to the generation of relatively shallow (~2.6 eV) IL traps (with a density on the order of 10^{19} cm^{-3}), which are characterized by a low relaxation energy upon electron capture (~0.36 eV), see Table I. Electrons can be transferred rather quickly (~10^{-6}sec @V_g=1.5V) through these defects, thus allowing for a high observed leakage current.

After SBD, the nature of the defects supporting the TAT current changes as follows from their energy characteristics, Table I. An exponential current increase after SBD, Fig. 5, is due to a much shorter electron tunneling distance through the unbroken portion of the dielectric to the states in the BD region (smaller x_T, ~ 0.35 nm, in Eq. (4), see also Fig. 4). This is accompanied by a larger λ (due to a smaller energy band offset, which is characteristic of thinner SiO_2 films) than the pre-SBD TAT current controlled by the electron transfer through the isolated, randomly distributed traps. An increase in the average relaxation energy of these traps and their deeper energy states results in higher current activation energy E_B, Table I, as calculated by Eq. (3). Overall, the characteristic time of the electron transfer through these traps, the density of which increased to ~10^{20} cm^{-3}, was reduced to ~10^{-11} sec at V_g=1.5V.

Table I. Extracted energy characteristics (in eV) of IL defects in 1.1nm SiO_2/3nm HfO_2 stack.

Parameters \ Degradation stage	Fresh	SBD	PBD
E_{rel}, relaxation energy	0.36	1.6	0.9
E_0, total energy differ.	-0.12	0.11	0.1
E_d, defect energy	2.6	3.1	3.1
E_B, activation energy	0.15	0.25	0.095

The PBD phase is associated with further propagation of the BD filament towards the substrate (further x_T decrease, $\leq$0.2 nm) and its higher overall oxygen deficiency. The latter is expected to lower the characteristic vibration frequency ω due to the less dense structure of the oxygen-depleted BD region, which, in turn, leads to a lower relaxation energy, $E_{relax} = S\,\hbar\omega$, and, subsequently, lower leakage current activation energy E_B, Eq. (3). However, as follows from the unchanged value of the defect energy E_d, the post-SBD evolution of the leakage current through the PBD phase is supported by the generation of similar defects, indicating that PBD is mostly a quantitative rather than qualitative evolution of SBD.

Leakage current activation energies at various phases of the stress-induced gate stack degradation calculated (Eq. (3)) using the trap parameters (Table I) are in excellent agreement with the experimental values, Fig. 6.

A proposed physical model describes the evolution of IL degradation through SBD and PBD as a propagation of a highly oxygen-deficient filament from the IL/high-k region toward the substrate. The model allows SILC temperature dependency and its exponential increase from fresh through PBD phases to be reproduced. Identifying the physical structure of the oxygen vacancy defects that support current through the BD filament in SiO_2, based on their energy characteristics will be addressed elsewhere.

4. V_{fb} Roll-Off Phenomenon

Among the most difficult challenges introducing metal electrodes into CMOS process are obtaining n- and n-type metal gate electrodes, with work functions (WFs) that match the Si valence and conduction band edges, respectively (as required to obtain low absolute threshold voltage values). It was found that when the gate stacks consisting of the metal electrode with the appropriate intrinsic WF, high-k dielectric, and interfacial SiO_2 layer were used in devices of practical interest, with scaled down SiO_2 layer, their EWF values were significantly less than those obtained in the test structures with thicker IL, most drastically in the gate stacks with high EWF (p-type electrodes) [21]. This phenomenon, which significantly limits the available options for metal/high-k transistor fabrication, is called flatband voltage roll-off (R-O).

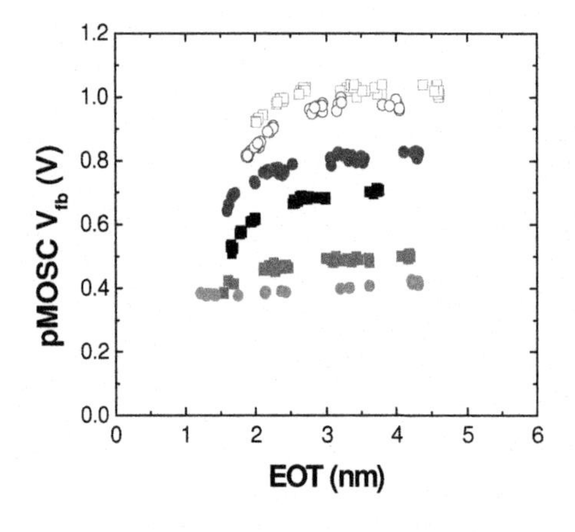

Fig. 9. Dependence of V_{fb} on EOT of the high-k terraced oxide capacitors with metal gates with different Work Functions.

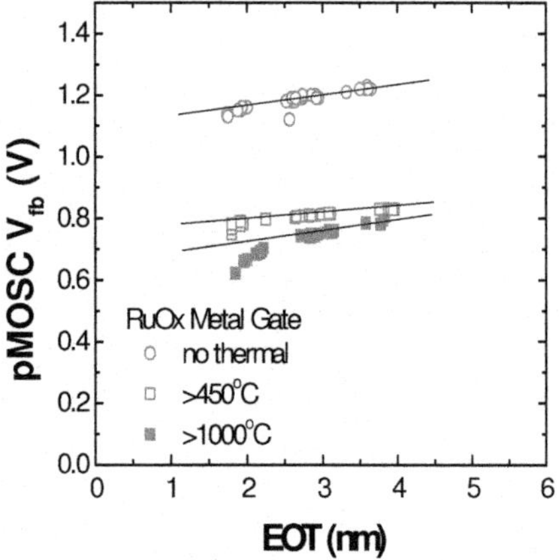

Fig. 10. V_{fb} roll-of (R-O) dependence on processing temperature in the 2 nm ALD HfO_2 capacitors with the RuO_2 gate electrode. EOT was changed by varying thickness of the underlying SiO_2 layer (terraced oxide capacitor structures [22]).

The following general dependencies of R-O characteristics were observed (to various degrees) in all devices with the metal/high-k/SiO_2 gate stacks:

• R-O increases when the electrode has a higher WF, and when high-k film is thicker. The effect of the electrode WF on R-O can be observed by comparing high-k gate stacks with different electrodes in Fig. 9. R-O onset shifts to the thicker interfacial SiO_2 in higher WF stacks and its magnitude is magnified by the high-k film thickness, as demonstrated in [22].

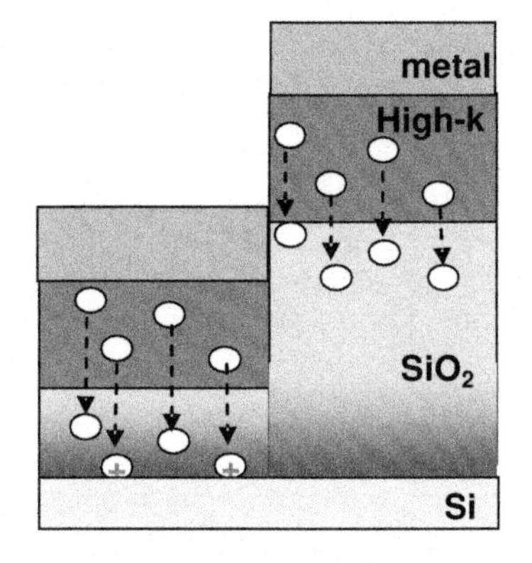

Fig. 11. Schematic of the proposed the R-O model. Shaded area represents strained transitional SiO_2 region.

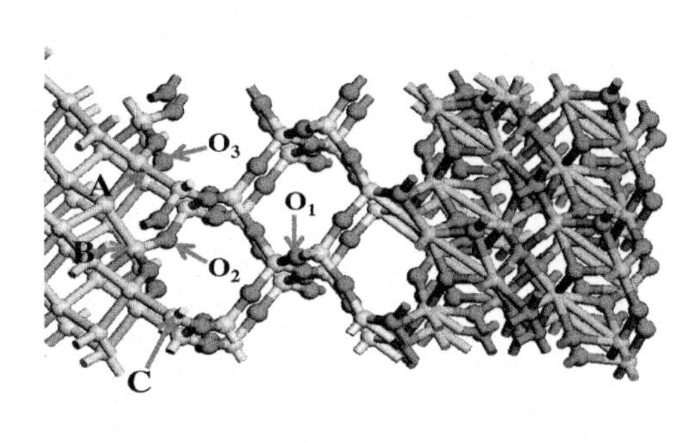

Fig. 12. Atomistic structure of the $Si/SiO_2/HfO_2$ stack. A, B, and C are Si atoms with +1, +2, and +3 oxidation states. O_1, O_2, and O_3 are removed oxygen atoms.

- R-O increases with a higher processing thermal budget (either higher temperatures and/or longer anneals (Fig. 10)).
- R-O is slightly greater on p-type substrates.

Mo_2N gated devices show that the R-O could be up to 100 mV more negative on p-type than on n-type substrates [23] although the substrate effect is less in mid-gap TiN devices.

4.1. *Roll-off mechanism*

Much of the above data points to the bottom SiO_2 interfacial layer as a region in the metal/high-k stack primarily responsible for R-O while other gate stack components may modulate, to a certain degree, its magnitude. We propose [24] that the R-O phenomenon as caused by enhanced positive charge generation within the interfacial SiO_2 layer when it becomes thinner than a certain critical value. These positive charges are suggested to be associated with the oxygen vacancies formed in SiO_2 due to its interaction with the overlaying high-k film/metal gate stack [15]. The generation of oxygen vacancies is expected to be significantly enhanced when O atoms are consumed from the highly strained transitional region in SiO_2 adjacent to the Si substrate [25]. Since a vacancy distribution through the SiO_2 thickness is oxygen diffusion-limited, enhanced vacancy generation due to oxygen consumption from the strained transitional SiO_2 layer is observed in only relatively thin SiO_2 films, when their thickness becomes comparable with the characteristic O diffusion length under given processing conditions. A final post-processed density gradient of O-vacancies in SiO_2 is affected by the processing temperature, SiO_2 density, high-k film thickness and composition, etc.

4.2. *Generation of positively charged defects*

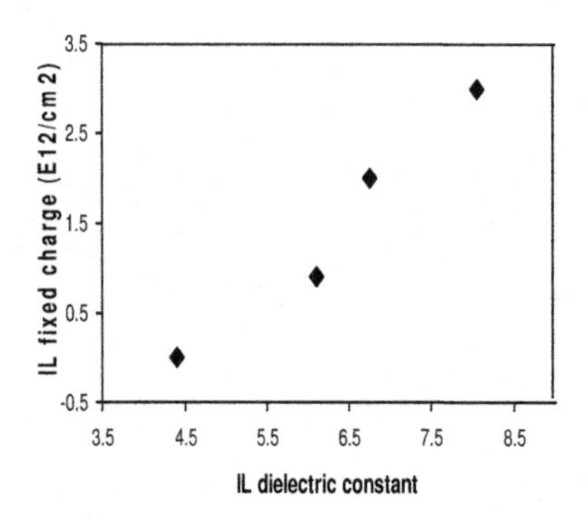

Fig. 13. Correlation between the dielectric constant and fixed positive charges in ~1nm interfacial SiO_2.

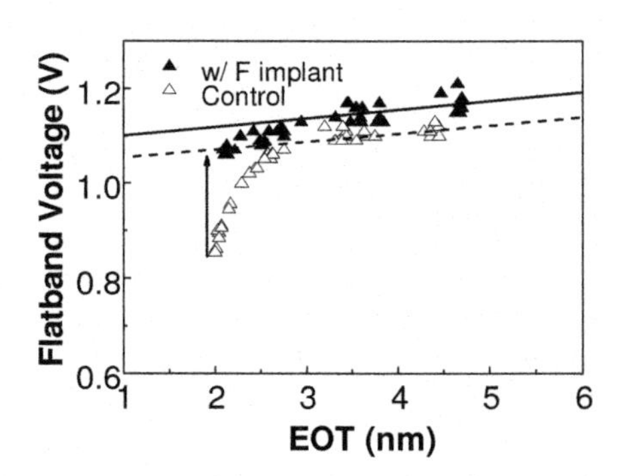

Fig. 14. R-O suppression by fluorine incorporated in the interfacial SiO_2 layer.

As was shown [26,27], Hf-based dielectrics consumes O from the interfacial SiO_2 layer due to a thermodynamic force that drives oxygen vacancies to migrate towards the SiO_2 layer. Therefore, the concentration of vacancies in SiO_2 is affected by their density in the adjacent high-k film, which effectively works as a source of vacancies. On the other hand, the calculations showed that for a high WF metal with a large oxide formation enthalpy, the dielectric/metal interface may be intrinsically unstable with respect to metal oxidation with simultaneous formation of oxygen vacancies in hafnia [28]. Thus, a higher density of oxygen vacancies in the interfacial SiO_2 layer in a high-k gate stack with higher WF metal electrodes should be expected.

To compare the formation energies of the oxygen vacancies in the SiO_2 bulk and transitional layer adjacent to the Si substrate, the $Si/SiO_2/HfO_2$ structure was modeled using density functional theory within the local density approximation (a plane wave code VASP); the total energy calculations were performed using the Vanderbilt-type ultra-soft pseudopotential method [29] (Fig. 12). It was found that formation energies for the vacancies at the O_2 and O_3 sites are lower than that of O_1 by 0.71 and 0.80 eV, respectively. For a doped substrate, two electrons left on the Si dangling bonds on the vacancy site can transfer to Si, thus forming a positively charged defect.

To verify the relationship of the positive charge at the dielectric/substrate interface and oxygen content, high-k transistors with varying stoichiometry of the interfacial SiO_2 layer (by changing the Si substrate treatment) were fabricated using a standard CMOS process [15]. After fabrication, all stacks were found to have approximately the same 1.1 nm SiO_2 layer physical thickness (by TEM) while k values of the post-processed SiO_2 layers (extracted using the EOT values of the total transistor gate stacks and of high-k/metal stacks) showed a strong dependence on the initial pre-high-k deposition conditions. Since the k value of the SiO_2 was shown to depend on the density of under-oxidized Si ions there, the observed increase of $k(SiO_2)$ indicates the SiO_2 layers in the corresponding gate stacks had a greater oxygen deficiency. On the other hand, by modeling the observed V_{fb} vs. equivalent oxide thickness (EOT) dependence obtained on these devices, the density of the positive charges in each of the stacks was extracted (Fig. 13). The data demonstrates a clear correlation between the degree of SiO_2 oxygen deficiency (as reflected by its k value) and the magnitude of positive charges in the SiO_2.

4.3. *Origin of R-O defects and R-O suppression*

Experimental data support the premises of the proposed R-O mechanism. Indeed, thicker high-k films function as a stronger source of oxygen vacancies in the interfacial SiO_2 layer leading to greater R-O. Similarly, higher WF electrodes, which generate oxygen vacancies in transition metal oxides more efficiently, are expected to enhance R-O by increasing the oxygen vacancy supply (through the higher-k film) to the interfacial oxide. Both generation and diffusion of oxygen vacancies in the dielectric stacks are controlled by the thermal budget leading to a strong R-O temperature dependency. Since the charge state of oxygen vacancies in the interfacial SiO_2 layer depends on the position of the substrate Fermi level, a positively charged vacancy state is more probable with p-substrates. Note that the R-O phenomenon may lead to misinterpretation of the mismatch of the measured and intrinsic WF values as caused by the Fermi-level pinning.

If R-O is caused by oxygen deficiency of the interfacial SiO_2 layer, then passivation of the oxygen vacancies should suppress R-O. Vacancies could be passivated, for instance, by incorporating either oxygen or fluorine into the interfacial layer. Indeed, implanting F^+ in the gate stack (1 nm SiO_2/2 nm HfSiO) followed by annealing at 1000°C/10 sec [30] significantly reduces R-O (Fig. 14) (post-processing SIMS data shows F accumulates at the SiO_2/Si interface). Similarly, R-O has been suppressed by employing a low-temperature (<450°C) oxygen annealing process [31], which is consistent with the model expectations.

Understanding of the R-O mechanism directs process improvement toward mitigating the generation of oxygen vacancies in the interfacial SiO_2 layer and suppressing their migration from the high-k dielectric into the interfacial oxide.

References

1. M. J. Kirton and M. J. Uren, *Advances in Physics* **38**, 367 (1989).
2. C. Leyris, F. Martinez, A. Hoffmann, M. Valenza, J. C. Vildeuil, *Microelectronics and Reliability* **47**, 41 (2007).
3. C. M. Chang, S. S. Chung, Y. S. Hsieh, L. W. Cheng, C. T. Tsai, G. H. Ma, S. C. Chien, S. W. Sun, The observation of trapping and detrapping effects in high-k gate dielectric MOSFETs by a new gate current Random Telegraph Noise (IG-RTN) approach, in *Proc. IEEE International Electron Devices Meeting*, San Francisco, CA, December 15-17, 2008 (IEDM 2008), p. 787.
4. S. Lee, H.-J. Cho, Y. Son, D. S. Lee, H. Shin, Characterization of Oxide Traps Leading to RTN in High-k and Metal Gate MOSFETs, Proc. IEEE International Electron Devices Meeting, Baltimore, MD, December 7-9, 2009 (IEDM 2009) p. 32.2.
5. M. B. Weissman, *Rev. Mod. Phys.* **60**, 537 (1988).
6. C. H. Henry and D. V. Lang, *Phys. Rev. B* **15**, 989 (1977).
7. Yu. E. Perlin, *Sov. Phys- Usp. (Usp Phys Nauk)* **6**, 542 (1964).

8. W. B. Fowler, J. K. Rudra, M. E. Zvanut, F. J. Feigl, *Phys. Rev. B* **41**, 8313 (1990).
9. A. Kimmel, P. Sushko, A. Shluger, and G. Bersuker, *ECS Trans.* **19**, 3 (2009).
10. D. Veksler, G. Bersuker, S. Rumyantsev, M. Shur, H. Park, C. Young, K. Y. Lim, W. Taylor, R. Jammy, Understanding noise measurements in MOSFETs: the role of traps structural relaxation, *Proc. IEEE International Reliability Physics Symposium*, Anaheim, CA, May 2-6, 2010 (IRPS 2010).
11. G. Bersuker, D. Heh, C. Young, H. Park, P. Khanal, L. Larcher, A. Padovani, P. Lenahan, J. Ryan, B. H. Lee, H.-H. Tseng, and R. Jammy, *IEEE IEDM Technical Digest*, 791 (2008).
12. S. Sahhaf, R. Degraeve, R. O'Connor, B. Kaczer, M. B. Zahid, P. J. Roussel, L. Pantisano, G. Groeseneken, Evidence of a new degradation mechanism in high-k dielectrics at elevated temperatures, *Proc. IEEE International Reliability Physics Symposium*, (IRPS 2009), p. 493.
13. G. Bersuker, N. Chowdhury, C. Young, D. Heh, D. Misra, R. Choi, Progressive Breakdown Characteristics of High-K/Metal Gate Stacks, *Proc. 45th annual IEEE Reliability Physics Symposium*, Phoenix, AZ, USA, 2007 (IRPS 2007) p. 49.
14. N. A. Chowdhury, G. Bersuker, C. Young, R. Choi, S. Krishnan, D. Misra, *Microelectronic Engineering* **85**, 27 (2008).
15. G. Bersuker, C. S. Park, J. Barnett, P. S. Lysaght, P. D. Kirsch, C. D. Young, R. Choi, B. H. Lee, B. Foran, K. van Benthem, S. J. Pennycook, P. M. Lenahan, and J. T. Ryan, *J. Appl. Phys.* **100**, 094108 (2006).
16. P. S. Lysaght, J. Barnett, G. I. Bersuker, J. C. Woicik, D. A. Fischer, B. Foran, H.-H. Tseng, and R. Jammy, *J. Appl. Phys.* **101**, 024105 (2007).
17. T. Ryan, P. M. Lenahan, G. Bersuker, and P. Lysaght, *Appl. Phys. Lett.* **90**, 173513 (2007).
18. X. Li, C. H. Tung, and K. L. Pey, Appl. Phys. Lett. 93, 262902 (2008); X. Li, C. H. Tung, K. L. Pey, V. L. Lo, The chemistry of gate dielectric breakdown, IEEE International Electron Devices Meeting, San Francisco, CA, 15-17 Dec. 2008 (IEDM 2008).
19. F. Giustino, A. Bongiorno, A. Pasquarello, *J. Phys.: Condens. Matter* **17**, S2065 (2005).
20. L. Larcher, IEEE Transactions on Electron Devices **50**, 1246 (2003).
21. B. H. Lee, J. Oh, H. H. Tseng, R. Jammy, H. Huff, *Materials Today* **9**, 32 (2006).
22. H.-C. Wen, R. Choi, G. A. Brown, T. Boscke, K. Matthews, H. R. Harris, K. Choi, H. N. Alshareef, H. Luan, G. Bersuker, P. Majhi, D. L. Kwong, B. H. Lee, *IEEE Electron Device Letters* **27**, 598 (2006).
23. J. K. Schaeffer, W. J. Taylor, S. B. Samavedam, D. C. Gilmer, D. H. Triyoso, R. I. Hegde, S. Kalpat, C. Capasso, M. Sadd, M. Stoker, A. Haggag, D. Roan and B. E. White Jr., Challenges for PMOS Metal Gate Electrodes and Solutions for Low Power Applications, *Proc. International Conference on Solid State Devices and Materials*, September 18-21, 2007, Ibaraki, Japan (SSDM 2007).
24. G. Bersuker, C. S. Park, H. C. Wen, K. Choi, O. Sharia, A. Demkov, Origin of the Flat-Band Voltage (Vfb) Roll-Off Phenomenon in Metal/High-K Gate Stacks, *Proc. 38th European Solid-State Device Research Conference*, 15 - 19 September 2008, Edinburgh (ESSDERC, 2008), p. 134.
25. Y. Harada, M. Niwa, T. Nagatomi and R. Shimizu, *Jap. Jour. of Appl. Phys.* **39**, 560 (2000).
26. W. L. Scopel, Antonio J. R. Dasila, W. Orellana and A. Fazzio, *Appl. Phys. Lett.* **84**, 1492 (2004).
27. N. Capron, P. Broqvist and Alfredo Pasquarello, *Appl. Phys. Lett.* **91**, 192905 (2007).
28. A. Demkov, *Phys. Rev. B* **74**, 085310 (2006).
29. D. Vanderbilt, *Phys. Rev. B* **41**, 7892 (1990).
30. Kisik Choi, Taeho Lee, J. Barnett, H. R. Harris, Seungsoo Kweon, C. Young, G. Bersuker, R. Choi, Seung Chul Song, Byoung Hun Lee, R. Jammy, Impact of Bottom Interfacial Layer on the Threshold Voltage and Device Reliability of Fluorine Incorporated PMOSFETS with High-K/Metal Gate, *Proc. 45th annual IEEE Reliability Physics Symposium*, Phoenix, AZ, USA, 2007 (IRPS 2007), p. 374.

31. C. S. Park, S. C. Song, C. Burham, H. B. Park, H. Niimi, B. S. Ju, J. Barnett, C. Y. Kang, P. Lysaght, G. Bersuker, R. Choi, H. K. Park, H. Hwang, B. H. Park, S. Kim, P. Kirsch, B. H. Lee and R. Jummy, Achieving Band Edge Effective Work Function of Gate First Metal Gate by Oxygen Anneal Processes: Low Temperature Oxygen Anneal (LTOA) and High Pressure Oxygen Anneal (HPOA) Processes, *Proc. International Conference on Solid State Devices and Materials*, September 18-21, 2007, Ibaraki, Japan (SSDM 2007).

ADVANCED SOLUTIONS FOR MOBILITY ENHANCEMENT IN SOI MOSFETS

L. PHAM-NGUYEN, C. FENOUILLET-BERANGER, P. PERREAU, S. DENORME,
G. GHIBAUDO, O. FAYNOT, T. SKOTNICKI, A. OHATA, M. CASSE, I. IONICA,
W. VAN DEN DAELE, K-H. PARK, S-J. CHANG, Y-H. BAE,
M. BAWEDIN, S. CRISTOLOVEANU

IMEP-LAHC (UMR 5130), Grenoble INP Minatec,
BP 257, 38016 Grenoble Cedex 1, France
sorin@enserg.fr

LETI, Grenoble, France
STMicroelectronics, Crolles, France

SOI technology offers ample room for scaling, performance improvement, and innovations. The current status is reviewed by focusing on several technological options for boosting the transport properties in SOI MOSFETs. The impact of series resistance, high-K dielectrics, and metal gate in advanced transistors is discussed. Carrier mobility measurements as a function of channel length and temperature reveal the beneficial effect of strain, mitigated however by various types of defects. The experimental data is exclusively collected from state-of-the-art, ultrathin body, fully depleted MOSFETs. Simple models are presented to clarify the mobility behavior.

Keywords: SOI; MOSFET; mobility; high-K; strain.

1. Introduction

Ten years ago, SOI promoters were hoping that this technology will soon penetrate the market place. These expectations were correct: many SOI chips are present in our servers, laptops, cars, watches, play stations, lighting systems, etc. SOI CMOS is indeed unchallenged when considering the combination of ultimate scalability, high speed and low power. In this paper, we will show that there is wide space for further improvements enabling a massive increase of the SOI segment in the global market of integrated circuits.

'More than Moore' strategies and technology developments are promising to drive us on a different avenue. 'Beyond CMOS' will offer sooner or later a variety of carbon nanotubes, graphene, spin and other innovative structures. But how will be the landscape looking like much later, beyond 'Beyond CMOS' era? CMOS will probably be the overwhelming technology and in very good shape. We believe that in 2035 our computers will still contain plenty of MOS transistors. Silicon and his friends (Ge, SiGe, GaN, etc) will not resign from competition in the micro-nano-electronics arena.

The real question is about the solutions available or necessary for further improving the MOSFET performance, functionality, and size. There are two knobs for tuning: electrostatics and transport.

The short-channel effects are governed by the gate-induced electrostatic properties of the structure. The gate control is reinforced if the transistor body is very thin [1]. This is why 'on insulator' structures are unavoidable. Since ultrathin SOI MOSFETs (Fig. 1d) are naturally fully depleted (FD), the best option is to keep the film undoped. A very thin buried oxide (BOX) and a ground plane underneath the BOX are valuable ingredients for scaling [2]. Together they reduce the penetration of fringing fields from source and drain into the body, via the substrate and BOX [1,2]. Ideal body control is actually achieved using multiple gates or a surrounding gate [3]. These aspects have been well documented recently and will not be developed in this paper.

Instead we will focus on the improvement of the transport properties. In the following sections, several boosters will be introduced and discussed based on experimental data collected in advanced SOI MOSFETs.

2. Series Resistance Reduction

Excessive series resistances jeopardize the performance of short-channel transistors. There is no point to implement solutions for improving the carrier mobility if the series resistance is not under control. Lowering the series resistance is the starting step for device optimization. This challenge is critical in ultrathin SOI where the sheet resistance ($R = \rho/t_{si}$) of source and drain is intrinsically large. In addition, the implantation process is less effective in very thin layers.

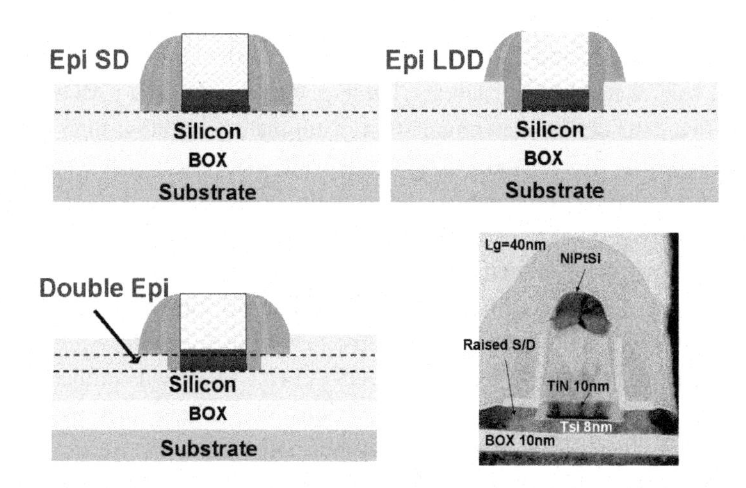

Fig. 1. Schematic configuration of SOI transistors with different solutions for selective epitaxial growth in source/drain regions: (a) Epi SD, (b) Epi LDD, and (c) Double Epi; (d) TEM image of fabricated transistor.

These considerations lead us to the processing of 'raised' source and drain terminals, using Selective Epitaxial Regrowth (SER). Figure 1 shows three possible variants:

- *Epi SD* – Epitaxial regrowth of source and drain from the nominal film thickness (8–10 nm) to a final junction thickness of 25–35 nm (Fig. 1a).
- *Epi LD*D – The epi-layer is grown after the formation of the offset spacer needed to protect the gate. The epi-layer now covers the LDD extension regions which are subsequently doped by tilted implantation (Fig. 1b). The second spacer is formed afterwards.
- *Double Epi* – After offset spacer formation and regrowth of LDD regions, a source/drain spacer is deposited and a second SER growth further raises the terminals (Fig. 1c).

The source/drain doping is activated by spike anneal for a few seconds. All these variants can greatly benefit from silicidation (NiPt) [4]. Our measurements indicate a significant lowering of the series resistance (~ 150 Ωμm), in particular for the Double-Epi silicided process.

3. High-K Dielectrics

High K dielectrics offer improved gate capacitance and reduced gate leakage current at the expense of a more defective interface with silicon. A detailed investigation of interface defects and optimization schemes was presented by Bersuker et al [5]. We focus on the carrier mobility behavior.

The devices were fabricated on 300 mm Unibond wafers with 70 nm Si film and 145 nm thick BOX. The transistor body was thinned down to 10 nm and left undoped. A 2.5 nm thick high-K dielectric was deposited by ALCVD. No significant difference between HfO_2 and HfSiON was observed.

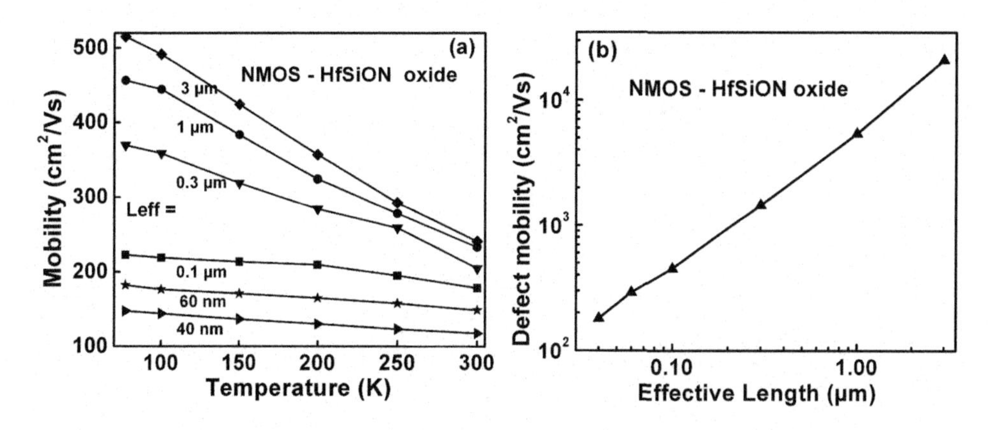

Fig. 2. (a) Comparison of front-channel low-field mobility variation as a function of temperature and gate length in n-channel SOI MOSFETs with HfSiON dielectric. (b) Mobility component related to neutral defect scattering versus channel length (300 K).

The low-field mobility was determined with the Y-function method: $Y = I_D/(g_m)^{0.5}$. Figure 2a shows typical results. In long-channel transistors, the electron mobility increases ($\mu \sim T^{-0.7}$) at low temperature, being dominated by acoustic phonon scattering ($\mu \sim T^{-1}$) with contribution from Coulomb scattering ($\mu \sim T$). By contrast, in short devices, the mobility $\mu(T)$ curve becomes nearly flat, independent on temperature. This difference points out the role of an additional scattering mechanism, which becomes more and more effective as the channel length shrinks. Using the Matthiessen rule, we have extracted in Fig. 2b, the corresponding mobility μ_{ND} as a function of channel length. This mobility component dominates the electron transport in short transistors and does not depend on temperature. It is attributed to the presence of neutral defects concentrated near the source/drain junctions. These 'edge' defects are generated during the source/drain implantation or the gate stack processing [4]. In long devices, most of the channel is free from the influence of the localized defects. But in short transistors, the defective 'edge' regions overlap leading to an increased density of defects which dramatically degrades the mobility.

An SOI transistor is a wonderful tool for comparing *in situ* the transport properties at the high-K and SiO_2 interfaces [4]. The diagnostic requires a single transistor where the current can flow at the front channel (Si-High K interface) or at the back channel (Si-SiO_2 interface). Figure 3a shows that the back channel mobility increases at low temperature, even in short transistors, being only marginally affected by the neutral defects. The localization of these neutral defects at the front interface suggests that the film-BOX interface may accommodate their recombination.

A consistent observation is the much higher mobility at the back channel compared with the front channel. At low temperature, the mobility ratio reaches a factor of two (Fig. 3a). It is clear that yet another scattering mechanism, inherent to the high-K interface, comes into play. The high-K mobility component μ_{HK} can be derived from the difference between the front and back channel reciprocal mobilities: $1/\mu_{HK} = 1/\mu_F - 1/\mu_B$ [4]. In short transistors, it is safe to include in the Matthiessen equation the mobility component related to neutral defects: $1/\mu_F = 1/\mu_B + 1/\mu_{HK} + 1/\mu_{ND}$ [6].

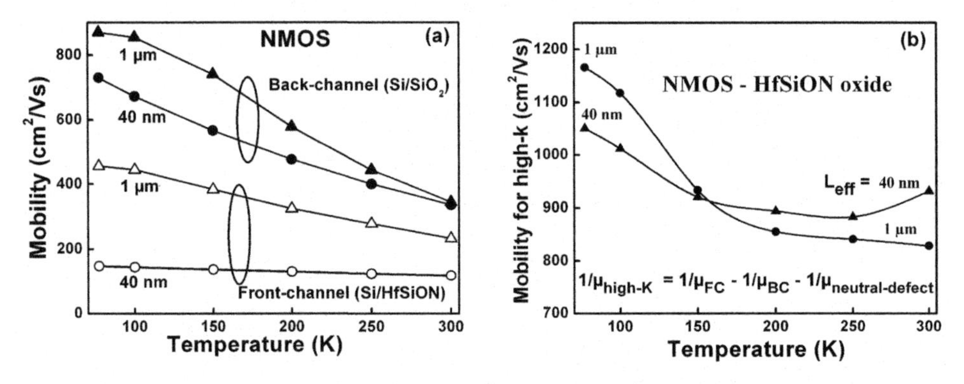

Fig. 3. (a) Electron mobility versus temperature at the front and back interfaces of long and short channel SOI MOSFETs. (b) Mobility component related to remote Coulomb centers, located in the high-K dielectric, versus temperature.

Figure 3b shows that the high-K mobility component does not depend on channel length and varies slowly with temperature. These features tend to confirm the prominent role of Coulomb scattering induced by remote charges located in the high-K dielectric or at its interface with the pedestal oxide [7]. Note that surface roughness scattering can be safely ignored as long as the mobility is measured at low vertical field.

4. Metal Gate

Midgap metal gates are necessary in FD SOI MOSFETs for suppressing the poly-depletion effect and adjusting simultaneously the threshold voltage of N- and P-channels. Gate optimization is conducted in terms of materials, thickness, and deposition techniques. Single-metal nitrides (TiN, TaN) are likely to induce nitrogen release which may potentially degrade the mobility [8].

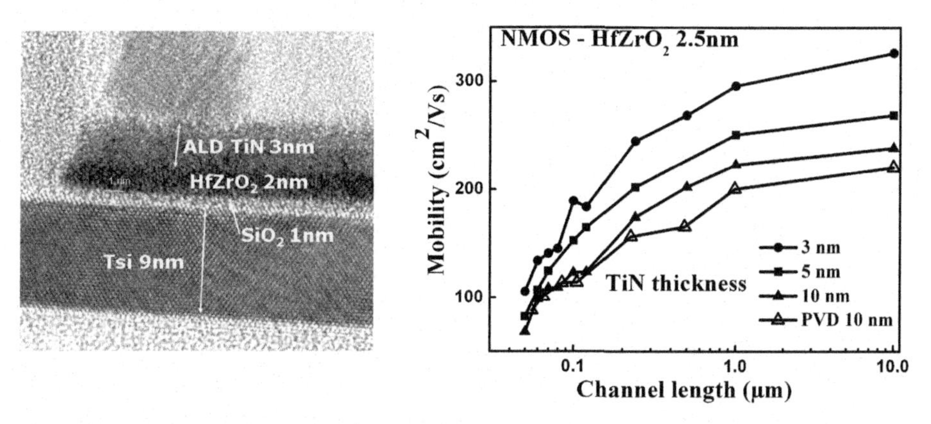

Fig. 4. (a) TEM image of the metal gate and high-K dielectric stack. (b) Electron mobility versus channel length for SOI MOSFETs with various metal gate thicknesses.

Our MOSFETs feature 9 nm thick film and 145 nm BOX. The gate stack is composed of 2 nm $HfZrO_2$ with 1.2 nm pedestal oxide, metal, and 80 nm polycrystalline Si (Fig. 4a). TiN and TaN metals with variable thickness (3–10 nm) were deposited by ALD or PVD methods. The various stacks do not modify the short-channel (L = 50 nm) electrostatic properties which are excellent: DIBL below 100 mV/V and subthreshold swing of 90 mV/decade (60 mV/decade in long channels). A thinner metal gate yields 30% improvement in electron mobility (see Figure 4b) and a marginal change in hole mobility [6].

As shown in Figure 5, the thinning of the metal gate from 10 nm down to 3 nm results in a threshold voltage shift of about 100 mV. There is no marked difference between TiN and TaN gates or between ALD and PVD depositions. Several mechanisms (variations of work-function, strain level, and trapping) were invoked to explain the V_T shift. Recent IPE (Internal Photoemission) experiments have revealed a change of the work-function with metal-gate thickness [9].

This change can result from: (i) the agglomeration of TiN islands at the metal-dielectric interface, (ii) the formation of interface dipoles, and (iii) charge trapping in the oxide due to the release and diffusion of nitrogen atoms. The latter scenario was confirmed by SIMS profiling. The practical conclusion is that the gate thickness adjustment can serve for threshold voltage tuning.

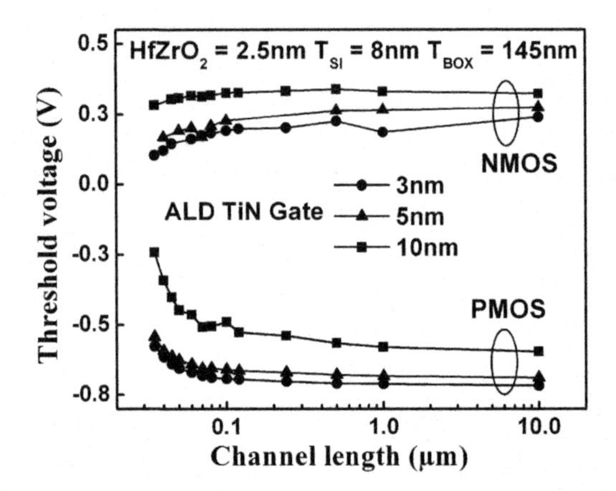

Fig. 5. Threshold voltage variation with channel length for various metal gate thicknesses in (a) N-channel and (b) P-channel SOI MOSFETs.

5. Strain

Strain engineering comes in multiple flavors. Biaxial strain, beneficial for both N- and P-channel MOSFETs, can be implemented within the SOI wafers using the Smart-Cut process [10]. The strained Si film is initially grown on relaxed SiGe layer, then is oxidized to form the BOX and, finally, is transferred on a support Si wafer. The oxide maintains the strain after the SiGe template layer is cut. The thinner the Si film, the higher the stress level. It is not perfectly clear what amount of wafer-level strain is lost during the processing of small area islands where MOSFETs are located.

Uniaxial strain should be compressive for holes and tensile for electrons. It can be introduced locally, at the transistor level, using dedicated stressors for each direction. *Longitudinal* strain is achieved by source-drain engineering: SiGe and SiC terminals are respectively used for P- and N-MOSFETs [11]. *Transversal* strain can be generated, essentially in narrow devices, using customized sidewall isolation techniques (STI, SiGe, etc). Finally, *vertical* strain is transferred from the gate stack and capping layers into the body. We will focus on the method exploiting the Contact Etch Stop Layer (CESL) [12].

The local strain in the transistor body depends on geometry and dimensions. In a large area device, regions with compressive and tensile strain can coexist, whereas in nanosize transistors the strain tends to become homogeneous.

According to the CESL composition and location (on top of the gate, on gate edges, on spacers, on extensions, etc), the strain effect can be modulated [13]. The strain is not uniform along the vertical direction. Its magnitude increases for ultrathin bodies and thicker CESL. The strain subsists after adding raised source/drain regions.

Our N- and P-channel MOSFETs (Fig. 1d) feature 1.6 GPa tensile and 3 GPa compressive strain, respectively. The fabrication recipe includes ultrathin Si film (8–10 nm), ALCVD deposited $HfZrO_2$ dielectric (2.5 nm), TiN gate, NiPt silicidation, and SiN-based CESL stressor.

Systematic measurements indicate that the strain does not modify the electrostatic behavior of short-channel devices. The threshold voltage roll-off, DIBL and subthreshold swing in N- and P- MOSFETs are hardly affected when switching from tensile to compressive strain [6]. By contrast, the transport properties and ON-current are strongly impacted in particular for p-channel transistors. Figure 6 illustrates the effect of compressive strain on the hole mobility. The mobility gain is modest for long channels but increases gradually to reach an impressive maximum (+80%) for 100 nm long P-MOSFETs.

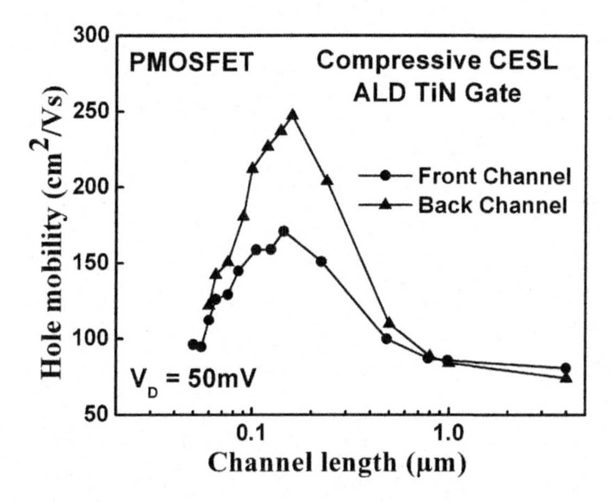

Fig. 6. Hole mobility versus channel length at the front and back channels of SOI MOSFETs with compressive CESL stressor.

The mobility evolution in Fig. 6 indicates the strain localization at the channel extremities. Detailed mechanical simulations of the device structure confirm that such 'stressed pocket' effect is maximized within 50 nm around the gate edge, at the corner between source/drain region and offset spacer [14,15]. As a matter of fact, the central region of a long channel (~1 μm) remains unstressed. But, in 30 to 100 nm long devices, the strain appears to be strong and rather uniformly distributed along the channel. Unfortunately, below 80–100 nm channel length, the hole mobility decreases dramatically. This degradation points out the onset of an adversely competing mechanism to stress, presumably neutral defect scattering [16].

An interesting aspect in Figure 6 is that the mobility gain is even more pronounced at the back channel. We deduce that the strain effect is transmitted from CESL through the whole film; it might even increase in the vertical direction, from the top to the bottom interface.

Mobility measurements at low temperature bring additional information. Figure 7a shows that in long channels the mobility increases steadily from 300 K to 77 K as a consequence of attenuated phonon scattering. The mobility behavior in short channels is rather unexpected. At room temperature, the hole mobility is much higher than in long devices due to the strain effect. At lower temperature, the mobility tends to saturate and the benefit of strain disappears.

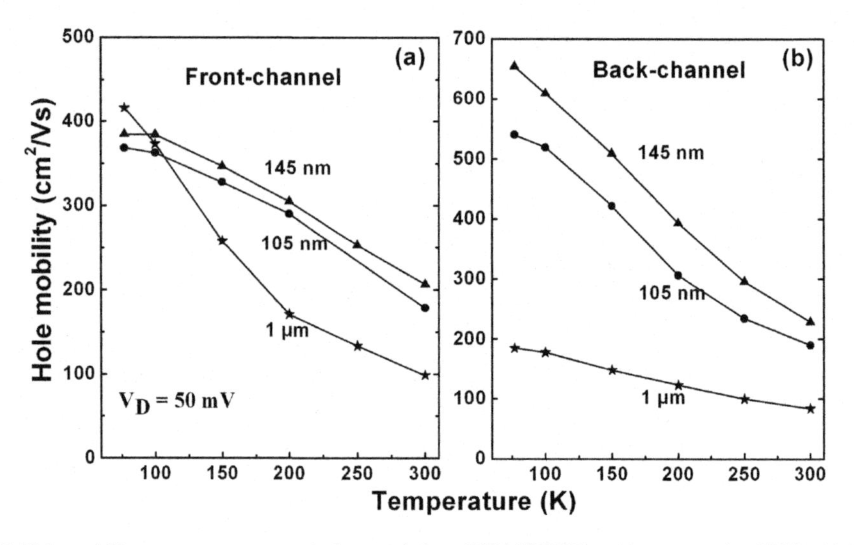

Fig. 7. Hole mobility versus temperature in long and short SOI MOSFETs with compressive CESL. (a) Front channel, (b) back channel.

The mobility dependence on temperature and gate length reveals the competition of several transport mechanisms in short devices. Indeed, the distribution of strain, neutral defects and scattering rate is highly inhomogeneous along the channel. These profiles do not affect the performance of long MOSFETs but become extremely effective in short transistors.

It is worth noting that the back-channel mobility (Fig. 7b) follows a more predictable evolution with temperature: in both short and long channels, the mobility increases substantially at low temperature, which implies less impact from neutral defects.

An analytical model has recently been proposed to explain the complex mobility behavior [6]. The model accounts for the stress tensor and its profile under the gate, from source to drain. Standard piezoresistive coefficients for silicon are used. It is found that the mobility starts to increase in channels shorter than 1 μm and tends to saturate below 0.1 μm.

For a more complete description of the hole mobility, several other contributions need to be accounted for [6]:

- *Phonon scattering*: μ_{ph} (x,T) = (300/T) μ_s (x), where μ_s (x) is the strain-dependent mobility at 300 K,
- *Neutral defect scattering*: μ_{ND} (x) = A_n/N_{ND} (x), where N_{ND} (x) is the longitudinal profile of the defect density, N_{ND} (x) = 4×10^{19} exp(-x/8nm) cm^{-3}, and A_n is a constant,
- *Remote Coulomb scattering* due to lateral depletion in the source/drain regions, next to the metallurgical junction: $1/\mu_{RCS}$ (x) = α G(x) Q_S, where Q_S is the space charge density in the depleted zones, $G(x) = [(1 + x/L_C)(1 + (L - x)/L_C)]^{-1}$ is the efficiency of remote Coulomb centers (ionized impurities), $L_C \approx 1.2$ nm, and $\alpha = 10^4$ Vs/C.

All above contributions are combined with the Matthiessen rule to provide a 'local' mobility value and a profile μ(x) along the channel. The 'global' mobility is obtained by integration of $1/\mu(x)$ from source to drain. This model successfully reproduces the experimental variation of hole mobility with both channel length (Fig. 6) and temperature (Fig. 7a). It is concluded that the compressive CESL strain boosts the hole mobility in short devices whereas neutral defects and remote Coulomb centers jeopardize the mobility gain in very short transistors.

An interesting avenue would be to combine the effects of various types of stressors (uniaxial and biaxial; vertical, lateral and longitudinal) and crystal orientations in order to achieve ideal transport conditions.

6. Novel Materials

Recent progress in SOI wafer fabrication is an additional source for device improvements. The idea is to break the happy marriage of Si and SiO_2 by introducing alternative materials. In principle, any combination of semiconductor film and buried dielectric, still part of the *Semiconductor* on Insulator SOI family, can be achieved by bonding and Smart-Cut process [10]. Germanium, Silicon Germanium and compound semiconductors (GaN, AsGa, etc) are envisioned for superior capability in terms of transport properties and optoelectronic functionalities.

The fabrication of Germanium on Insulator (GeOI) substrates is already well controlled [17]. An original method, proposed by Tezuka [18], is the Ge condensation which exploits the selective oxidation of a SiGe layer, with low Ge content, grown on SOI. During oxidation, only Si atoms are consumed whereas the diffusion of Ge atoms is blocked by the BOX barrier. This process results in a thinner, Ge-enriched film. The initial SOI wafer can actually be totally converted into a 10 nm thick GeOI film. When needed, the layer thickness can be completed by Ge epitaxy.

Experimental results show that the hole mobility increases with Ge content (Fig. 8), reaching excellent values far beyond Si capabilities (200 cm^2/Vs in 10 nm film, 400 cm^2/Vs in 50–100 nm film, and even 700 cm^2/Vs for strained GeOI) [19].

This appreciable improvement for holes is balanced by a clear degradation in electron mobility, probably due to a peculiar distribution of traps in the band gap [20]. An attractive solution is the co-integration of N-channel SOI and P-channel GeOI transistors within the same chip. Local Ge condensation in selected islands enabled the fabrication of SOI–GeOI hybrid substrates and corresponding devices [21].

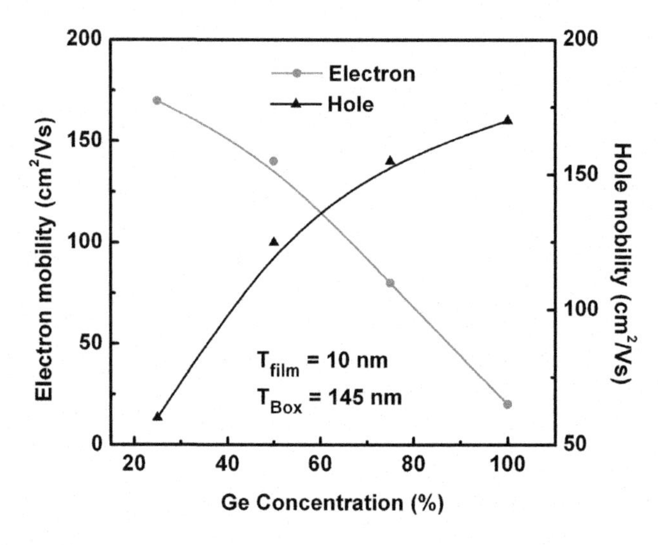

Fig. 8. Electron and hole mobilities in ultrathin (10 nm) GeOI layers versus Ge content. The measurements were performed directly on as-fabricated wafers using the Pseudo-MOSFET technique (after Nguyen et al [19]).

A parallel avenue is to change the BOX material with the aim of solving the problem of transistor self-heating. The heat dissipation in the silicon substrate is blocked by the poor thermal conductivity of SiO_2. Buried alumina, Si or Al nitrides, and other dielectrics have been positively evaluated for possible replacement [22]. Silicon on Diamond (SOD) wafers were recently demonstrated using a three-step process: (i) growth of diamond on Si, (ii) bonding to a Si substrate, and (iii) removal of the unnecessary portion of the stack [23]. SOD MOSFETs show excellent characteristics, without self-heating or mobility degradation.

An interesting concept is to promote the BOX as an active layer. For example, nitride BOX can induce strain in the film whereas a transparent BOX (quartz, glass, diamond) opens the door to photonic applications. A buried ONO stack is useful as a reservoir of nonvolatile charges, enabling the new paradigm of 'unified' SOI memory. The idea is to use a single SOI transistor for both volatile and non-volatile functions.

The capacitorless DRAM mode is emulated by temporary accumulation of majority carriers in the floating body and the flash mode by 'permanent' carrier charging in the ONO layer [24].

Direct wafer bonding is also capable to mate circuits pre-processed on a variety of substrates. This technique can be expanded for the fabrication of 3D stacked circuits.

7. Conclusions

SOI technology is gaining momentum as bulk CMOS is confronted to scaling, power dissipation and performance challenges. The milestones are measurable in nanometers, for the transistor feature size, and in billions for the number of transistors in a System-on-a-Chip. SOI is able to take CMOS far beyond its current frontiers because the transistor miniaturization is more comfortable 'on Insulator'.

We argue that fully depleted SOI MOSFETs are ideal candidates to address most of the CMOS challenges in a successful and cost effective manner. Their performance can be boosted by using advanced methods (high-K/metal-gate stack, strain, raised source/drain, ultrathin nanometer-size film, thin BOX) documented in this paper.

Engineered strain effects are beneficial but very complex, with a 3D profile within the transistor body. Maximum strain is achieved longitudinally and laterally near the gate edges. Technology optimization is critical for preventing the strain-induced gain to be competed by the presence of additional defects, generated by implantation and high-K/metal-gate stacks. We have shown that these defects are responsible for special scattering mechanisms which affect the carrier mobility in very short and thin channels.

Future research is needed to achieve the 'perfect' transistor which will combine wafer level strain, 3D process-enabled strain, hybrid channel/wafer orientations, multiple gates, and alternative materials.

Acknowledgements

This work has been supported by various projects including EUROSOI+, NANOSIL, and WCU of KOSEF. Our many SOI colleagues are deeply thanked.

References

1. T. Ernst, C. Tinella, C. Raynaud and S. Cristoloveanu, Fringing fields in sub-0.1μm fully depleted SOI MOSFETs: optimization of the device architecture, Solid-State Electronics, **46**, 373–378 (2002).
2. C. Fenouillet-Beranger, S. Denorme, P. Perreau, C. Buj, O. Faynot, F. Andrieu, L. Tosti, S. Barnola, T. Salvetat, X. Garros, M. Casse, F. Allain, N. Loubet, L. Pham-Nguyen, E. Deloffre, M. Gros-Jean, R. Beneyton, C. Laviron, M. Marin, C. Leyris, S. Haendler, F. Leverd, P. Gouraud, P. Scheiblin, L. Clement, R. Pantel, S. Deleonibus and T. Skotnicki, FDSOI devices with thin BOX and ground plane integration for 32nm node and below, Proc. ESSDERC, 206–209 (2008).
3. S. Cristoloveanu, Silicon on Insulator Technology, *The VLSI Handbook*, 2nd Edition, 4.1–4.23, W-K. Chen ed. (CRC Press, Boca Raton, USA, 2006).
4. L. Pham-Nguyen, C. Fenouillet-Beranger, A. Vandooren, T. Skotnicki, G. Ghibaudo and S. Cristoloveanu, In-situ comparison of Si/High-K and Si/SiO_2 channels properties in SOI MOSFETs, IEEE Electron Device Letts. **30**(10), 1075–1077 (2009).

5. G. Bersuker, C. S. Park, J. Barnett, P. S. Lysaght, P. D. Kirsch, C. D. Young, R. Choi, B. H. Lee, B. Foran, K. van Benthem, S. J. Pennycook, P. M. Lenahan and J. T. Ryan, *J. Appl. Phys.* **100**, 094108 (2006).
6. L. Pham-Nguyen, C. Fenouillet-Beranger, G. Ghibaudo, T. Skotnicki and S. Cristoloveanu, Mobility enhancement by CESL strain in short-channel ultrathin SOI MOSFETs, Solid-State Electronics, **54**(2), 123–130 (2010).
7. S. Barraud, O. Bonno and M. Cassé, The influence of Coulomb centers located in HfO_2/SiO_2 gate stacks on the effective electron mobility, J. Appl. Phys. **104**, 073725 (2008).
8. C. Fenouillet-Beranger, P. Perreau, L. Pham-Nguyen, S. Denorme, F. Andrieu, L. Tosti, L. Brevard et al, Hybrid FDSOI/bulk high-k/metal gate platform for low power (LP) multimedia technology, Proc. IEDM, 1–4 (2009).
9. M. Charbonnier, C. Leroux, V. Cosnier, P. Besson, E. Martinez, N. Benedetto, C. Licitra, N. Rochat, C. Gaumer, K. Kaja, G. Ghibaudo, F. Martin and G. Reimbold, Measurement of dipoles/roll-off /work functions by coupling CV and IPE and study of their dependence on fabrication process, IEEE Trans. Electron Devices, in press (2010).
10. S. Cristoloveanu and G. K. Celler, SOI materials and devices, *Handbook of Semiconductor Manufacturing Technology*, 4.1–4.52, 2nd Edition, Y. Nishi and R. Doering eds. (CRC Press, London, 2007).
11. D. Chanemougame, S. Monfray, F. Boeuf, A. Talbot, N. Loubet, F. Payet, V. Fiori, S. Orain, F. Leverd, D. Delille, B. Duriez, A. Souifi, D. Dutartre and T. Skotnicki, Performance boost of scaled Si PMOS through novel SiGe stressor for HP CMOS, VLSI Tech. Digest, 180–181 (2005).
12. G. Eneman, P. Verheyen, A. D. Keersgieter, M. Jurczak and F. De Meyer, Scalability of Stress Induced by Contact-Etch-Stop Layers: A Simulation Study, IEEE Trans. Electron Devices, **54**(6) 1446–1453 (2007).
13. C. Gallon, C. Fenouillet-Beranger, S. Denorme, F. Boeuf, V. Fiori, N. Loubet, A. Vandooren et al, Mechanical and electrical analysis of a strained liner effect in 35 nm FDSOI devices with ultra-thin silicon channels, Jap. J. Appl. Phys. **45**(4B), 3058–3063 (2006).
14. C. Ortolland, S. Orain, J. Rosa, P. Morin, F. Arnaud, M. Woo, A. Poncet and P. Stolk, Electrical characterization and mechanical modeling of process induced strain in 65 nm CMOS technology, Proc. ESSDERC, 137–140 (2004).
15. F. Payet, F. Boeuf, C. Ortolland and T. Skotnicki, Nonuniform mobility enhancement techniques and their impact on device performance, IEEE Trans. Electron Devices, **55**(4), 1050–1057 (2008).
16. A. Cros, K. Romanjek, D. Fleury, S. Harrison, R. Cerruti, P. Coronel, B. Dumont et al, Unexpected mobility degradation for very short devices: A new challenge for CMOS scaling, Proc. IEDM, 439–442 (2006).
17. C. Deguet, C. Morales, J. Dechamps, J. M. Hartmann, A. M. Charvet, H. Moriceau, F. Chieux, A. Beaumont, L. Clavelier, V. Loup, N. Kemevez, G. Raskin, C. Richtarch, F. Allibert, F. Letertre and C. Mazure, Germanium-on-Insulator structure realized by the Smart-Cut^{TM} technology, IEEE International SOI Conf., 96–97 (2004).
18. T. Tezuka, S. Nakaharai, Y. Moriyama, N. Sugiyama and S. Takagi, Selectively-formed high mobility SiGe-on-Onsulator pMOSFETs with Ge-rich strained surface channels using local condensation technique, VLSI Symp. Tech. Dig. 198–199 (2004).
19. Q. T. Nguyen, J.-F. Damlencourt, B. Vincent, L. Clavelier, Y. Morand, P. Gentil and S. Cristoloveanu, High quality Germanium-on-insulator wafers with excellent hole mobility, Solid-State Electronics, **51**(9), 1172–1179 (2007).
20. W. Van Den Daele, E. Augendre, C. Le Royer, J.-F. Damlencourt, B. Grandchamps and S. Cristoloveanu, Low-temperature characterization and modeling of advanced GeOI pMOSFETs: mobility mechanism and origin of the parasitic conduction, Solid-State Electronics, **54**(2), 205–212 (2010).

21. C. Le Royer, J.-F. Damlencourt, K. Romanjek, Y. Lecunff, H. Grampeix, Y. Mazzocchi, V. Carron, F. Nemouchi, C. Arvet, C. Tabone, M. Vinet, L. Hutin, P. Batude and L. Clavelier, High mobility CMOS: first demonstration of planar GeOI pFETs and SOI nFETs, Proc. EUROSOI 2010, 21-22 (2010).
22. N. Bresson, S. Cristoloveanu, C. Mazuré, F. Letertre and H. Iwai, Integration of buried insulators with high thermal conductivity in SOI MOSFETs: thermal properties and short channel effects, Solid-State Electronics, **49**(9), 1522–1528 (2005).
23. J.-P. Mazellier, J. Widiez, F. Andrieu, M. Lions, S. Saada, M. Hasegawa, K. Tsugawa et al, First demonstration of heat dissipation improvement in CMOS technology using silicon-on-diamond (SOD) substrates, IEEE Int. SOI Conf., 1–2 (2009).
24. K.-H. Park, S. Cristoloveanu, M. Bawedin, Y.-H. Bae, K.-I Na and J.-H. Lee, Double-gate 1T-DRAM cell using nonvolatile memory function for improved performance, Solid-State Electronics, in press (2010).

ELECTRON SCATTERING IN BURIED InGaAs/HIGH-K MOS CHANNELS

S. OKTYABRSKY, P. NAGAIAH, V. TOKRANOV, M. YAKIMOV, R. KAMBHAMPATI
and S. KOVESHNIKOV

College of Nanoscale Science and Engineering, University at Albany-SUNY, NY 12203, USA

D. VEKSLER, N. GOEL and G. BERSUKER

International SEMATECH, Albany, NY 12203, USA

Hall electron mobility in buried QW InGaAs channels, grown on InP substrates with HfO_2 gate oxide, is analyzed experimentally and theoretically as a function of top barrier thickness and composition, carrier density, and temperature. Temperature slope α in $\mu \sim T^{\alpha}$ dependence is changing from α=-1.1 to +1 with the reduction of the top barrier thickness indicating the dominant role of remote Coulomb scattering (RCS) in interface-related contribution to mobility degradation. Insertion of low-k SiO_x interface layer formed by oxidation of thin *in-situ* MBE grown amorphous Si passivation layer has been found to improve the channel mobility, but at the expense of increased EOT. This mobility improvement is also consistent with dominant role of RCS. We were able to a obtain a reasonable match between experiment and simple theory of the RCS assuming the density of charges at the high-k/barrier interface to be in the range of $(2\text{-}4)\times10^{13}$ cm^{-2}.

Keywords: MOSFET; A_3B_5; InGaAs; high-k oxide; mobility.

1. Introduction

As CMOS scaling continues, novel channel materials such as III-V compound semiconductors with high-k gate oxides are considered to improve channel transport, intrinsic gate delay and to reduce power consumption of MOSFETs [1]. In order to keep high mobility in the channel, the channel carriers have to be separated from the primary scattering sources, i.e. the interface with high-k oxide, and thus buried quantum well (QW) channel is a promising option, although the top semiconductor barrier adds up to the equivalent oxide thickness (EOT) and thus degrades scaling properties of the transistors. Therefore, the rates of decay of different scattering mechanisms related to a high-k oxide interface with top barrier thickness are of great importance.

Another motivation of this study is the effect of interface passivation layers (IPL) on scattering. Various IPLs with relatively low dielectric constant were proposed to improve interface properties [2], most recently amorphous Si [3] and Ge [4], which are partially oxidized after deposition/annealing of high-k oxides. Similar to a Si/high-k oxide interface which always contains low-k SiO_x interlayer [5], the a-Si IPL improves mobility in InGaAs channels [6]. We will show, however, that this improvement is obtained with a trade-off of an increased EOT.

Finally, mobility is typically evaluated from electrical characteristics of MOSFETs, but in III-V's with relatively high interface trap density and source-drain contact resistance, determination of the channel charge and hence the channel mobility is quite a challenging problem. The mobility, however, serves as a good metrics of interface quality. Hall measurements of structures with semiconductor/high-k interface is, therefore, a good alternative method to characterize the properties of this interface, in particular with respect to charge responsible for scattering. In this paper, with the goal to establish the baseline for mobility vs. EOT optimization, we study the scattering mechanisms in 2D gas due to the presence of high-k oxide and a-Si interface passivation layer.

2. Experimental Details

The samples were grown on semi-insulating Fe-doped InP(001) substrates using multi-chamber molecular-beam epitaxy (MBE) system. The epi-structures included 400nm-thick undoped $In_{0.52}Al_{0.48}As$ buffer layer lattice-matched to the substrate a 10nm-thick compressively strained $In_{0.77}Ga_{0.23}As$ quantum well (QW) channel, and a top semiconductor barrier layers with different thicknesses up to 50 nm. The top barrier layers with thickness ≤3nm had $In_{0.53}Ga_{0.47}As$ composition, and the thicker barriers consisted of $In_{0.52}Al_{0.48}As$ capped with 2 monolayers of $In_{0.53}Ga_{0.47}As$ to prevent oxidation of Al-containing compound. Following the growth of III-V layers, a 10nm-thick high-k HfO_2 gate oxide layer was grown in-situ by reactive electron beam evaporation of metallic Hf in oxygen at a pressure of 10^{-6} Torr. Samples with interface passivation layer (IPL) contained thin (0.5-2nm) amorphous Si layer deposited in-situ by MBE. Deposition of high-k oxide and further annealing resulted in partial oxidation of about 0.5nm of the IPL [7,8]. Samples were annealed at 500-600 ^{0}C in an RTA furnace under nitrogen flow. TEM cross-sectional images and layout of the structures without and with a-Si IPL are shown in Fig. 1(a,b).

In order to compare buried QW channel transport with different thicknesses of the top barrier layer, we have chosen bulk modulation doping scheme below the QW channel, separated with a 5nm-thick undoped spacer layer to reduce ionized impurity scattering. Electron transport properties of the structures were measured using Van der-Pauw method (Fig. 1c) from 77K to room temperature (RT), Hall measurements were performed in the magnetic field of 0.5-1 T. CV methods supported with transmission electron microscopy were used to analyze interface structure and chemistry of the gate stack.

3. Mobility Calculation

The electron mobility dependence versus both, the temperature and the distance between InGaAs channel and HfO_2 layer can be calculated assuming the major electron scattering mechanisms to be remote Coulomb scattering (RCS) and lattice vibrations. We employed the approach of Barraud et al. [5] to calculate the RCS limited mobility. The calculations

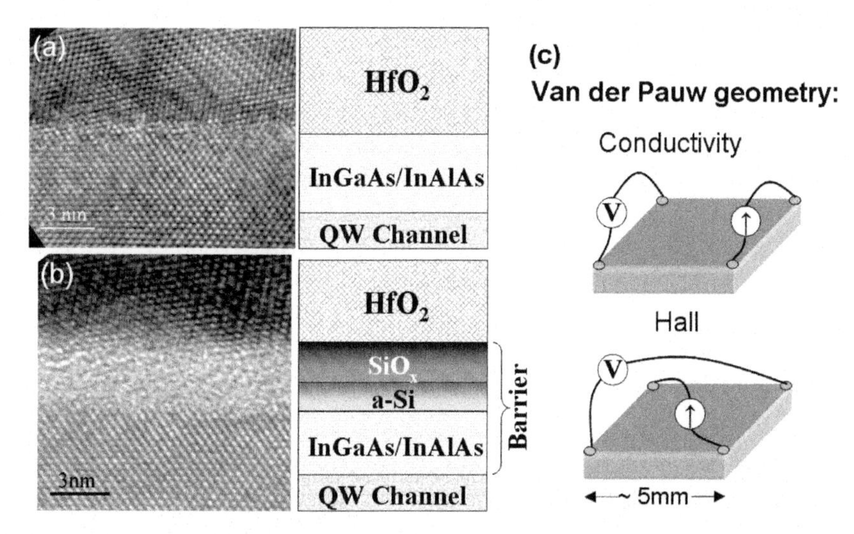

Fig. 1. TEM micrographs of interface regions of the gate stacks without (a) and with (b) a-Si IPL. (c) Van der Pauw sample geometry for conductivity and Hall-effect measurements.

are based on the solution of the Poisson equation in a multi-layer structure with the layers having different dielectric constants. Screening of the scattering potential is taken into account using the Yokoyama and Hess formula [9], which adequately reproduce screening effects for both non-degenerate and the degenerate 2D cases. For simplicity, we restricted ourselves to the case of a single parabolic sub-band and replaced the electron wave function in the channel with the δ-function in the direction perpendicular to the interface.

The mobility dependences on the top barrier thickness and temperature were calculated for heterostructures with a III-V barrier (it is assumed that the entire barrier is a single layer with the dielectric constant k=14) and for the structure with the top part of the barrier replaced with a SiO_x layer (k=3.9). The total mobility was obtained from the calculated μ_{RCS} using the Mathiessen's rule:

$$\frac{1}{\mu}=\frac{1}{\mu_{RCS}}+\frac{1}{\mu_{Bulk}} \tag{1}$$

where the value for the mobility in the "bulk" QW (mostly phonon-limited) was obtained form the data measured on the samples with 50 nm thick top barrier [10]:

$$\mu_{Bulk}=11000\cdot(T/300K)^{-1.1}\{cm^2/Vs\} \tag{2}$$

It should be also mentioned that the model used assumes additive contribution of all the charged traps in the dielectric to the electrons scattering. In fact, it is true only in case of moderate density of charges ($N_C<d^{-2}$, with d - a separation of interface charges from the maximum of electron density in QWs), while in case of large density of traps

their contribution to scattering is partially reduced. Also, presence of both positive and negative charges in the HfO_2 would cause the scattering potential to be dipole-like, leading to a stronger dependence of the electron scattering rate vs. the barrier thickness [11]. In the mobility calculation (Eq. 1) we have neglected other mechanisms of oxide/interface-related scattering, such as surface roughness and soft-phonon scattering that can contribute to mobility. The reason for this is a very pronounced temperature dependence of experimental interface-related mobility characteristic of RCS as will be shown below. Consideration of other mechanisms would naturally improve the theory-experiment match.

4. Results and Discussion

Fig. 2 shows the effect of the top semiconductor barrier thickness on electron Hall mobility in 10nm-thick compressively strained $In_{0.77}Ga_{0.23}As$ QW channel. The mobility is reduced from about 11000 cm^2/Vs in a reference sample with thick (50nm) barrier and the same modulation doping, to about 1500 cm^2/Vs in a surface QW channel. Many samples with different doping concentration and annealing temperatures were analyzed [6]. The mobility maximizes after annealing at 500-600 ^0C and at sheet carrier density close to $2x10^{12}$ cm^{-2} and reduces at lower and higher carrier densities [10]. Mobility reduction at low carrier density is consistent with reduced screening of the RCS by the channel electrons, and at high density with conduction band nonparabolicity and filling of the high-effective-mass upper valleys.

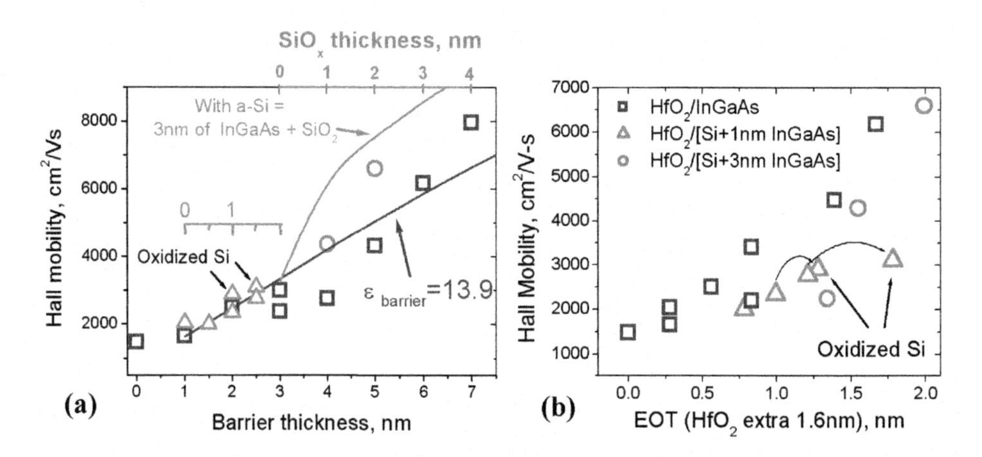

Fig. 2. Room temperature Hall electron mobility in $In_{0.77}Ga_{0.23}As$ QW channel as a function of top barrier layer thickness. (a) Comparison of experimental (symbols) and theoretical (lines) mobility dependences on physical thickness of top barrier layer, which consists of 0-7 nm semiconductor layer (squares), 3 nm- (circles) or 1nm-thick (triangles) InGaAs semiconductor barrier with Si interlayer partially or entirely oxidized. (b) Hall mobility vs. equivalent oxide thickness (EOT) for stacks with different top barriers as in (a). Electron sheet density is in the range of $(2-4)x10^{12}$ cm^{-2}.

Insertion of a low-k SiO_x interface layer formed by oxidation of a thin *in-situ* MBE grown amorphous Si IPL noticeably improves the channel mobility (compare in Fig. 2a mobility of ~4000 cm^2/Vs for a 5 nm thick III-V barrier with mobility of ~7000 cm^2/Vs for 3 nm thick III-V barrier on top of 2 nm thick SiO_x barrier, i.e., for the same barrier physical thickness of 5 nm). This is qualitatively similar to the mobility improvement due to interface SiO_x layer in Si/HfO_2 gate stack due to damping of the remote Coulomb [5,12] or soft phonon [13] scattering from the high-k oxide. An interesting feature of the low-k interlayer is an effective screening of the scattering potential. The screening increases with reduction of the dielectric constant of an interlayer, or more accurately, with the increased difference in the dielectric constants of high-k and low-k layers.

To verify the effect of Si IPL induced screening, we have grown two sets of gate stacks consisting of 1nm-thick top InGaAs semiconductor barrier and in-situ a-Si IPLs with 1 and 1.5 nm thicknesses. Further on, one subset of the samples were oxidized at 100 ^{0}C in the air and then returned to the MBE system for high-k deposition. Another subset was directly transferred into high-k deposition chamber under UHV conditions. In this case, we expect only about 0.5nm of the Si IPL to be oxidized [7,8], while the Si IPLs in the air-exposed of samples are likely to be entirely oxidized, and the dielectric constant of the interface layer is reduced from that of amorphous Si (about 9) to close to that of SiO_2 (k=3.9). The two samples with entirely oxidized IPL have shown the increased mobility (indicated by arrows in Fig. 2) as compared to the samples with partially oxidized IPL.

Fig. 2(b) contains the same set of mobility data plotted against equivalent oxide thickness of the top barrier (including semiconductor barrier and IPL). The dielectric constant of the Si IPL (partially or entirely oxidized) is calculated using the values specified above. When the IPL is used, the mobility improvement comes at the expense of too large EOT. In fact, thicker top semiconductor barrier has a superior effect on mobility than low-k SiO_x interlayer.

The solid lines in the Fig. 2(a) show mobility calculated using the approach described in Section 3. The phonon-limited mobility of 11000 cm^2/Vs is used from measurements of deeply buried QW channels. The RCS-limited mobility is obtained with the same charge distribution in the 4nm-thick HfO_2 layer for both types of structures with and without the SiO_x interlayer, totaling $4x10^{13}$ cm^{-2}. It is worth noting that the standard frequency dependent conductance-voltage (G-V) methods usually would detect the traps located up to a few angstroms deep in the dielectric, i.e. primarily the interfacial states. Therefore, the trapped charge density determined from the G-V measurements might be several orders of magnitude smaller than the density of charges contributing to the mobility degradation in the devices under study. The calculations match the experimental results quite reasonably given the simplicity of the used approach and a single fitting parameter varied in the simulation. The calculated mobility higher than experimental one in the samples with Si IPL is likely due to the higher dielectric constant of the formed SiO_x interlayer than that of stoichiometric SiO_2 employed in the calculations.

Temperature dependence of mobility is shown in Fig. 3. The samples with deeply buried (*d*=50 nm) QW channel shows $\mu \sim T^{-1.1}$ trend, which is close to $\mu \sim T^{-1.2}$ typically observed in n-type InGaAs QWs in the 80-300K temperature range [14] with the main contribution from polar optical phonon scattering and some effect of almost temperature-independent alloy disorder and hetero-interface roughness scattering. The samples with thinner top barriers show the reduced slope of the curves and eventually reversed slopes in the samples with d<3nm. More accurately, positive μ(T) slope was observed in the samples with RT mobility below ~3000 cm^2/Vs and is close to $\mu \sim T^{+1.0}$ in the surface QW channels indicating the dominant role of remote Coulomb scattering (RCS) due to charges in the oxide and at the interface. The net positive μ(T) slope in QWs with relatively high electron density (~$2x10^{12}$ cm^{-2}) is a quite unexpected result that indicates considerably stronger contribution of scattering due to the HfO_2 interface with InGaAs than that with Si [12,13,15].

The calculated mobility vs. temperature curves are shown in Fig. 3 by open symbols and considering RCS and bulk phonon scattering measured on the samples with 50 nm thick top barrier. All the curves were obtained with a single fitting parameter, namely the density of trapped charges at the high-k/barrier interface, which was set to $2x10^{13}$ cm^{-2}.

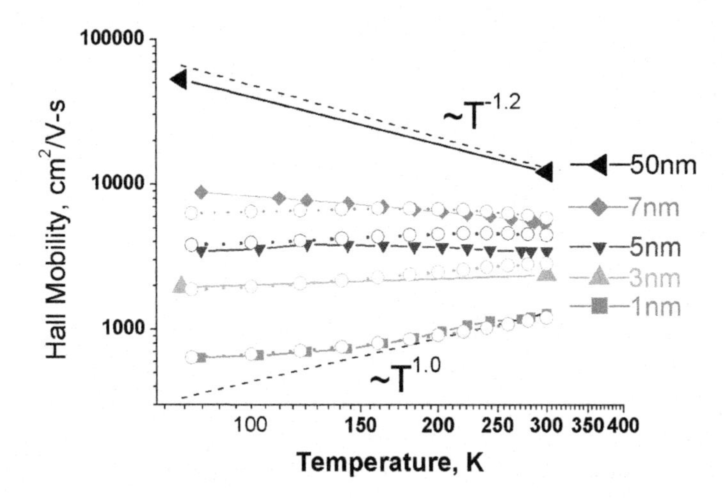

Fig. 3. Temperature dependence of Hall mobility in QWs at various top barrier layer thicknesses: experimental (solid symbols) and theoretical (open symbols) curves. Slopes for optical phonon ($T^{-1.2}$) and Coulomb scattering ($T^{+1.0}$) are indicated by dashed line. Electron sheet density is close to $2x10^{12}$ cm^{-2} in all the samples and does not change significantly with temperature.

To analyze the μ(T) dependence in the samples with Si IPL, we plot the interface-limited mobility vs. temperature in Fig. 4. The interface-limited mobility is obtained from the measured values using the equation in the Fig.4. Similarly to the gate stacks without SI IPL, the slope of mobility is positive for thin barriers and flattens out when the top barrier thickness increases. The slope values in the $\mu_{Int}(T)$ dependences are plotted vs. EOT in Fig. 5 for two set of samples with and without Si IPL. The slope reduction is very similar in these samples indicating that the RCS is the dominant interface-related scattering mechanisms in both cases. These similarities also suggest that the charges responsible for scattering are located in HfO_2, or more precisely, close to the HfO_2 interface and have similar densities. The effect of the InGaAs/SiO_x interface on mobility is, therefore, quite minute. It is also confirmed by significantly increased mobility of QW channels without HfO_2. For example, the channel with 3 nm InGaAs barrier and 2nm SiO_x on top and without high-k oxide has demonstrated mobility of 7550 cm^2/Vs (corresponding to interface-limited mobility of 24000 cm^2/Vs) at electron sheet density of 1.7×10^{12} cm^{-2} [10].

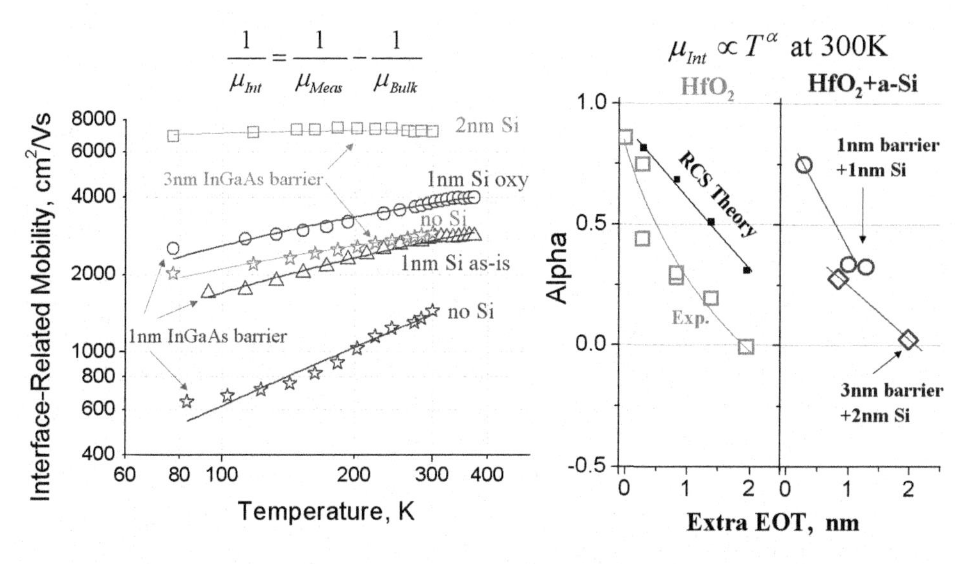

Fig. 4. Temperature dependence of interface-related Hall mobility in QWs with various top barrier layers and 10 nm HfO_2.

Fig. 5. EOT dependence of the exponent in $\mu \sim T^{\alpha}$ relationship at room temperature for two sets of samples with and without Si IPL and 10 nm HfO_2. Theoretical dependence is shown by solid circles.

The Fig. 5 also shows the $\mu(T)$ slope dependence on EOT calculated using the theoretical approach discussed above. The graph just illustrates the trend of this dependence, but also indicates the difficulty of accurate treatment of the RCS, in particular, evaluation of temperature dependence of 2D screening. The similarity of the experimentally observed and calculated trends provides an additional evidence that the RCS is the dominant interface-related mechanism of scattering.

Another important question is the origin of the charges contributing to scattering. Recently, we have estimated the interface trap density from the concentration dependence in QWs vs. annealing [6] on the same structures, and from the conductance measurements on similar gate structures with Ni gate metal, both giving $D_{it} \sim 10^{13}$ cm^{-2}-eV^{-1} in the upper part of the bandgap. Interface states at the semiconductor surface can contribute to the RCS once they are charged. Since the neutrality surface level in InGaAs with high In content (on InP substrate) stays close to the conduction band, the interface trapped charge is relatively low, and in our samples hardly exceeds low 10^{12} cm^{-2}. On the other hand, our calculations require (2-4) x10^{13} cm^{-2} to account for the observed mobility degradation. Moreover, it is well known that remote charge of the order of 10^{12} cm^{-2} does not contribute significantly to channel mobility degradation in high-electron mobility transistors (HEMTs) with modulation doping. HEMTs with spacer thickness (layer between the channel and doping ions) of 5nm may show mobilities over 10^6 cm^2/Vs at cryogenic temperatures proving negligible contribution of RCS [11].

The results on the structures with Si IPL also indicate that the interface traps are not contributing to scattering. Firstly, we have found, that contrary to GaAs and low-In InGaAs, the a-Si IPL does not decrease D_{it} at $In_{0.53}Ga_{0.47}As$/high-k oxide interface [16]. The major improvement that Si IPL provides for the latter stack is a significantly better thermal stability, but the reduction of D_{it} is minute. Secondly, as we have shown above, the interface-limited scattering is mostly due to the charge located at SiO_x/HfO_2 interface rather than at InGaAs/SiO_x interface. The SiO_x/HfO_2 interface, as well as the defects in HfO_2 are not capable of recharging or recharge at very long time scale, and can hardly contribute to interface traps observed typically through the conductance or charge pumping methods at relatively high frequencies. Thus, we can conclude that fixed charges in HfO_2 are mostly responsible for the observed RCS effects. It is also important that these charges might be of both signs or might constitute fixed dipoles, thus reducing the shift in flat-band or threshold voltages in these gate stacks. In fact, the charge density required to account for the observed RCS is very high and would cause a noticeable band bending if combined from the charges of the same sign. However, typically band bending does not exceed 0.2-0.3V, indicating that total fixed charge at the interface is below ~2x10^{12} cm^{-2} for typical oxide capacitance of 1μF/cm^2.

5. Conclusions

We have analyzed Hall electron mobility in buried QW InGaAs channels with HfO_2 gate oxide as a function of top barrier thickness and composition, carrier density, and temperature. Temperature slope α in $\mu \sim T^{\alpha}$ dependence is changing from α=-1.1 to +1 with the reduction of the top barrier thickness indicating the dominant role of remote Coulomb scattering (RCS) in interface-related contribution to mobility. Insertion of the low-k SiO_x interlayer results in higher mobility as compared to the III-V barrier of the same physical thickness, the contribution of the SiO_x layer to EOT is higher too. Based on the mobility dependence on temperature and barrier thickness/composition, the dominant role of RCS was demonstrated. We were able to obtain a reasonable match between experiment and simple theory of the RCS assuming rather large charge density (2-4x10^{13}cm^{-2}) in the gate stack. These charges are likely the fixed charges of both signs and dipoles located within a few nm in HfO_2.

Acknowledgement

The work was supported by INTEL Corporation and SRC Focus Center Research Program as a part of MSD Center.

References

[1] R. Chau, S. Datta, and A. Majumdar, IEEE Compound Semiconductor Integrated Circuit Symposium Tech. Digest, 17, 2005.

[2] S. Oktyabrsky, M. Yakimov, V. Tokranov, R. Kambhampati, H. Bakhru, S. Koveshnikov, W. Tsai, F. Zhu, and J. Lee, Int. J. High Speed Electronics and Systems **18,** 761 (2009).

[3] S. Koveshnikov, W. Tsai, I. Ok, J. Lee, V. Tokranov, M. Yakimov, and S. Oktyabrsky, Appl. Phys. Lett. **88,** 022106 (2006).

[4] H.-S. Kim, I. Ok, M. Zhang, T. Lee, F. Zhu, L. Yu, J. C. Lee, S. Koveshnikov, W. Tsai, V. Tokranov, M. Yakimov, and S. Oktyabrsky, Applied Physics Letters **89,** 222904 (2006).

[5] S. Barraud, O. Bonno, and M. Casse, J. Appl. Phys. **104,** 073725 (2008).

[6] S. Oktyabrsky, P. Nagaiah, V. Tokranov, S. Koveshnikov, M. Yakimov, R. Kambhampati, R. Moore, and W. Tsai, MRS Proc. **1155,** C02 (2009).

[7] S. Oktyabrsky, V. Tokranov, M. Yakimov, R. Moore, S. Koveshnikov, W. Tsai, F. Zhu, and J. C. Lee, Materials Sci. Eng. B **135,** 272 (2006).

[8] R. Kambhampati, S. Koveshnikov, V. Tokranov, M. Yakimov, R. Moore, W. Tsai, and S. Oktyabrsky, ECS Trans. **11,** 431 (2007).

[9] K. Yokoyama and K. Hess, Phys. Rev. B **33,** 5595 (1986).

[10] P. Nagaiah, V. Tokranov, M. Yakimov, S. Oktyabrsky, S. Koveshnikov, W. Tsai, D. Veksler, and G. Bersuker, J. Crystal Growth (2010) (in press).

[11] K. Lee, M. S. Shur, T. J. Drummond, and H. Morkoc, Journal of Applied Physics **54,** 6432 (1983).

[12] S. Barraud, L. Thevenod, M. Casse, O. Bonno, and M. Mouis, Microelectronic Engineering **84,** 2404 (2007).

[13] K. Maitra, M. M. Frank, V. Narayanan, V. Misra, and E. A. Cartier, Journal of Applied Physics **102,** 114507 (2007).

[14] T. Matsuoka, E. Kobayashi, K. Taniguchi, C. Hamaguchi, and S. Sasa, Jap. J. Appl. Phys., Part 1 **29,** 2017 (1990).

[15] M. A. Negara, K. Cherkaoui, P. Majhi, C. D. Young, W. Tsai, D. Bauza, G. Ghibaudo, and P. K. Hurley, Microelectronic Engineering **84,** 1874 (2007).

[16] R. Kambhampati, S. Koveshnikov, V. Tokranov, M. Yakimov, T. Heeg, M. Warusawithana, D. G. Schlom, W. Tsai, and S. Oktyabrsky, 2009 6th Int. Symp. Advanced Gate Stack Technology (ISAGST), Tech. Digest (2009).

LOW FREQUENCY NOISE AND INTERFACE DENSITY OF TRAPS IN InGaAs MOSFETs WITH $GdScO_3$ HIGH-K DIELECTRIC

S. RUMYANTSEV

Rensselaer Polytechnic Institute, Troy NY 12180-3590, USA and
Ioffe Institute of Russian Academy of Sciences, 194021 St. Petersburg, Russia
roumis2@rpi.edu

W. STILLMAN and M. SHUR

Rensselaer Polytechnic Institute, Troy NY 12180-3590, USA

T. HEEG and D.G. SCHLOM

Department of Materials Science and Engineering, Cornell University, Ithaca, NY 14853

S. KOVESHNIKOV, R. KAMBHAMPATI, V. TOKRANOV and S. OKTYABRSKY

College of Nanoscale Science and Engineering, University at Albany-SUNY, NY 12203, USA

Insulated gate n-channel enhancement mode InGaAs field effect transistors with the $GdScO_3$ high-k dielectric have been fabricated and studied. The low frequency noise was high indicating a high interface density of traps. Trap density and its dependence on the gate voltage have been extracted from the noise and conductance measurements.

Keywords: MOSFET; A_3B_5; InGaAs; high-k; noise; trap density.

1. Introduction

InGaAs MOSFETs are the prime candidates to replace Si MOSFETs for low-power high-speed digital and microwave analog applications. Recent progress in high-k dielectrics has allowed for their use in aggressively scaled InGaAs MOSFETs. These devices have potential advantages of high electron mobility, extremely low gate leakage current and great potential for further scaling.

Ternary rare-earth oxides such as $LaAlO_3$, $GdScO_3$, and $LaLuO_3$ are promising for application as gate dielectrics because of their high dielectric constants, large bandgaps and band offsets, amorphous structure and thermal stability up to 1000°C. However in spite of the progress in high-k technology, even Si MOSFETs with high-k dielectric still suffer from high interface trap density affecting the device characteristics and leading to high levels of the low frequency noise.

In this work, we report on electrical and noise characteristics of n-channel enhancement mode InGaAs MOSFETs with the $GdScO_3$ gate dielectric providing insight into mechanisms of the low frequency noise.

2. Experimental Details

Growth of the III-V structures was carried out on p-type InP (001) substrates using molecular beam epitaxy (MBE). The samples consisted of a 150 nm thick carbon doped ($2x10^{18}$ cm^{-3}) buffer layer followed by an 80 nm thick $In_{0.53}Ga_{0.47}As$ layer doped with carbon to a concentration of $3x10^{17}$ cm^{-3}. Passivation of the semiconductor surface was done using an in-situ MBE grown 0.5 nm thick *a*-Si layer [1]. The surface was further capped with a dense As_2 layer to prevent surface oxidation while samples are transferred to another MBE system where the arsenic capping layer was thermally desorbed and $GdScO_3$ was grown by co-deposition of Gd and Sc in oxygen ambient at a pressure of 10^{-6} Torr with the ratio between the metallic elements maintained close to unity. The final $GdScO_3$ thickness was varied between 7 and 15nm. MOSFETs were fabricated using a 200 nm thick TaN gate metal, Si+ (50 keV energy and dose of 10^{14} cm^{-2}) ion implantation into the source and drain regions, implant activation at 750ºC for 15sec, and PdGe ohmic contacts [2].

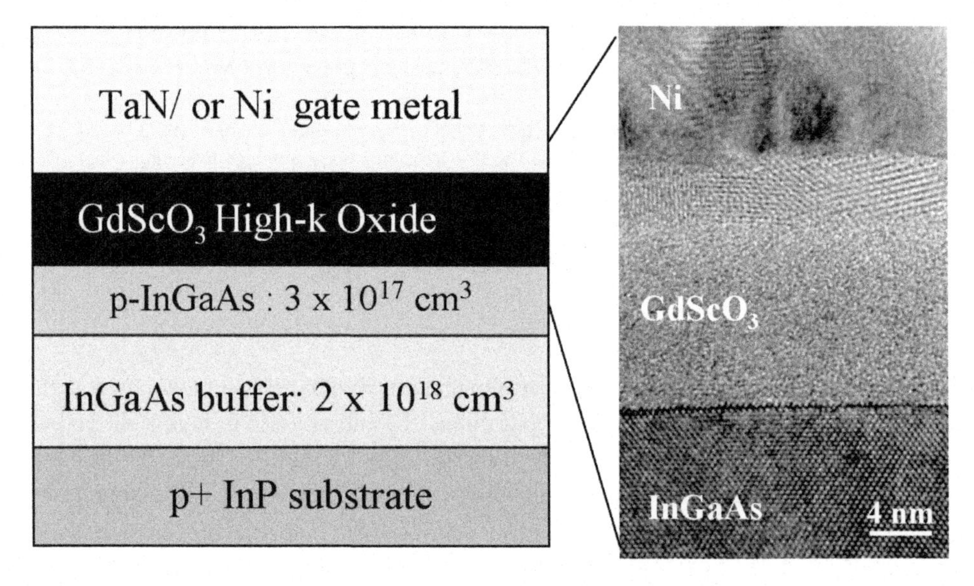

Fig. 1. Cross section and TEM image of InGaAs/InP MOS structure after 800^0C annealing. The interlayer between the metal and high-k oxide is a catalytically crystallized Gd_2O_3 phase.

The layers structure of the device and its transmission electron microscope (TEM) images are shown in Fig. 1. $GdScO_3$ seems to have crystallized at the metal-oxide interface. Figure 2 shows the cross section view of the transistor structure and top view photograph of the transistors configuration.

The low-frequency noise was measured in a frequency range from 1 Hz to 50 kHz at 300 K using an electro-statically shielded probe station with 10-μm diameter tungsten probes under controlled pressure. The transistors were biased in the linear region in the common source mode. The voltage fluctuations S_V from the load resistor R_L connected in series with the drain were analyzed by a SR770 FFT Spectrum Analyzer.

The spectral noise density of the short circuit drain current fluctuations, S_I, was calculated using as $S_I = S_V\left[(R_L + R_{SD})/(R_L R_{SD})\right]^2$, where R_{SD} is the drain-to-source resistance.

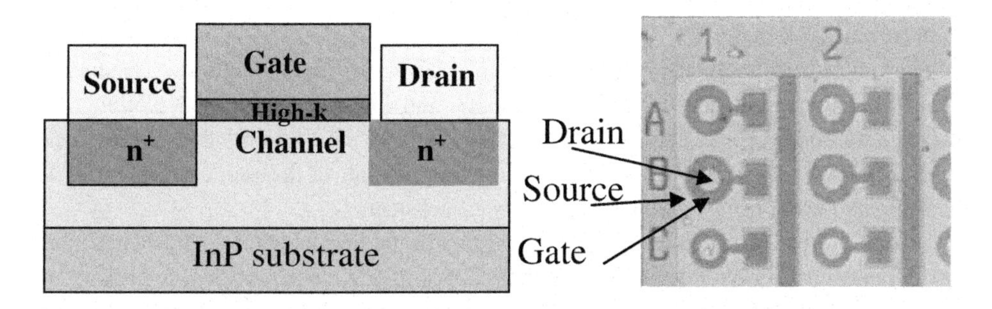

Fig. 2. Cross section and top views of the transistors structure.

3. Results and Discussion

Figure 3 shows the current voltage characteristics of one of the transistors with a gate dielectric thickness t=15nm.

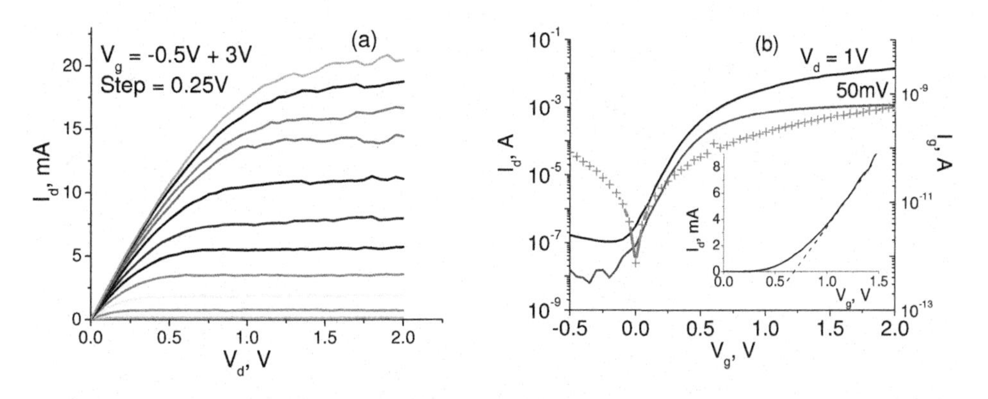

Fig. 3. Output (a) and input (b) current voltage characteristics of the transistor with gate dielectric thickness t=15nm and W/L=37. Crosses in Fig. 3b shows the gate leakage current, which is orders of magnitude smaller than the drain current. The inset shows input current voltage characteristic for V_d=1V in the linear scale. Dashed line shows linear approximation used to define the threshold voltage.

The output characteristics (Fig. 3a) demonstrated good linearity at small drain voltage, V_d, and very good saturation at $V_d>(V_g-V_t)$. These MOSFETs had a threshold voltage of $V_t = 0.4$-0.9V (depending on the $GdScO_3$ thickness), the ideality factor for the sub-threshold slope $\eta \approx 2$-2.3, and on-to-off ratio of ~ 10^5 (see Fig. 3b). For the given substrate doping level and gate capacitance the subthreshold swing was estimated to be $S \approx 70$mV/decade (see [3], p.447). The actual subthreshold swing of $S \approx 130$mV/decade is explained by a high interface trap density close to the conduction band minimum. Estimation of trap density from the subthreshold swing yields $D_{it} \approx C_g (q / SkT \ln 10 - 1) = 4\times10^{12} cm^{-2} eV^{-1}$ (for S=130mv/decade). The gate leakage current, in spite of the relatively large gate area of the test transistors (~8.8×10^{-5}cm^2), was several orders of magnitude smaller than the drain current at all biases.

Capacitance-voltage characteristics (Fig. 4) show extremely low frequency dispersion in accumulation of 2.2% per decade, and low stretch-out in depletion. The interface trap density was decreasing down to 2×10^{12}cm^{-2}eV^{-1} towards the valence band (see the inset of Fig. 4) and was quite high in the upper half of the gap, corresponding to the estimated value from subthreshold slope measurements.

Mobility was estimated from the drain current in the linear region (at V_d=50mV) and carrier concentration obtained from integration of split CV characteristics. A peak value of about 2000 cm2/Vs was obtained at inversion carrier density 3×10^{11} cm^{-2} and dropped to 1200cm^2/Vs at 3×10^{12} cm^{-2}.

Low frequency noise spectra were measured over a wide range of the gate voltage, V_g, and at the drain voltage V_d=50-100mV. The noise spectra were close to the $1/f^{\gamma}$ dependence with γ=0.9-1 over the entire range of frequencies, f, and the V_g and V_d values for all transistors (examples of the noise spectra are shown in Fig. 5)

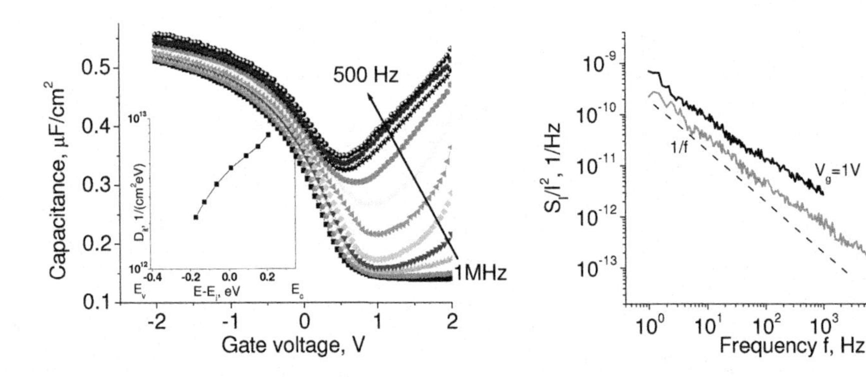

Fig. 4. Capacitance-voltage characteristics of the MOS-capacitors with 15 nm $GdScO_3$ dielectric and Ni gate metal measured at 500Hz-1MHz frequency range. Inset shows D_{it} extracted using conductance method.

Fig. 5. Two examples of the noise spectra for the transistor with gate dielectric thickness t=15nm. V_d=10mV.

The spectral noise density S_I measured at constant gate voltage was proportional to I_d^2 across the wide range of I_d which is typical for linear resistors and field effect transistors with good quality drain and source contacts. On the other hand, the dependence of noise on V_g had a form, which differed qualitatively from that for conventional silicon MOSFETs.

For Si MOSFETs as well as for many other FETs, the noise S_I/I_d^2 decreases in strong inversion $\sim 1/(V_g-V_t)^2$ (or even steeper if the contact resistance is high). As seen from Fig. 6a, this is not the case for transistors under investigation: S_I/I_d^2 depends on the gate voltage as $S_I/I_d^2 \sim 1/(V_g-V_t)$ for high values of (V_g-V_t) and tends to saturate at small values of (V_g-V_t).

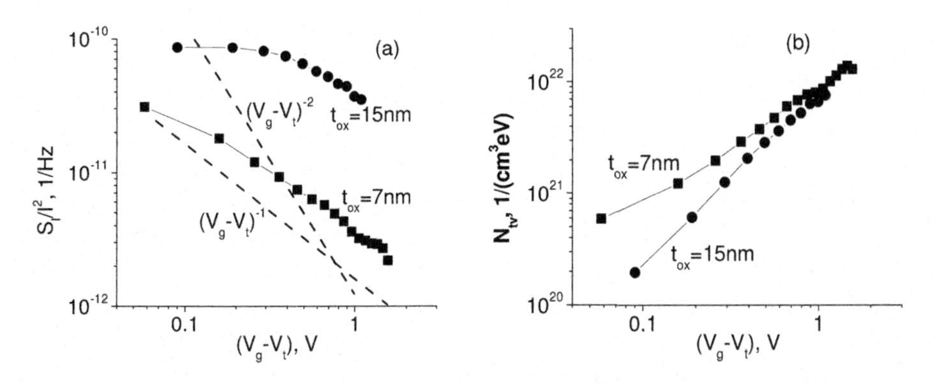

Fig. 6. (a): Spectral noise density S_I/I^2 at f=10Hz versus (V_g-V_t) for two transistors with the gate dielectric thickness t_{ox} =7nm and t_{ox} =15nm. (b): Trap density calculated for the same transistors as a function of (V_g-V_t).

Similar dependences were observed in amorphous and polycrystalline TFTs [4-7] and in SiC MOSFETs [8]. Such dependencies can be explained by the energy dependence of the density of traps responsible for noise.

The effective volume trap density in the oxide, N_{tv}, which is responsible for the $1/f$ noise, can be estimated using the McWhorter model [9,10]:

$$\frac{S_I}{I_d^2} = \frac{\lambda k T N_{tv}}{f W L_g n_s^2}. \quad (1)$$

Here $\lambda \approx 10^{-8}$cm is the tunneling constant, N_{tv} is the volume trap concentration in the oxide in $(\text{eV cm}^3)^{-1}$, and

$$n_s = \frac{C_g \eta k T}{q^2} \ln\left[1 + \exp\frac{q(V_g - V_t)}{\eta k T}\right] \quad (2)$$

is the sheet electron concentration in the channel. Equation (2) describes the gate voltage dependence of concentration both below and above the threshold voltage [11]. Using eq. (2) instead of simplified $n_s = C_g(V_g - V_t)/q$ usually describes well the noise behavior in weak inversion and below threshold [12,13] (see [14,15] for the detailed analysis of noise in subthreshold).

Figure 6b shows the trap density N_{tv} calculated using eq. (1) for transistors with two different dielectric thicknesses. As seen, the trap density is very high and increases with the gate voltage. In spite of the difference in the noise amplitude for the transistors with the different oxide thicknesses, the trap density for these two types of devices is of the same order of magnitude. The McWhorter model and eq. (1) yield the trap density on the Fermi level. Therefore the increase of the trap density with increasing $(V_g\text{-}V_t)$ points out to an increase of the density of state traps with energy increasing towards the conduction band.

The low frequency noise in MOSFETs is often interpreted based on the unified noise model which includes correlated mobility fluctuations [16-18]. Even though this mechanism might contribute to noise, a high concentration of traps shows that trapping should play a dominant role in noise behavior.

The noise analysis in MOSFETs and other devices is also often based on the empirical Hooge relation, which postulates the noise to be inversely proportional to the total number of carriers in the sample. However, the Hooge equation does not reveal the noise mechanism.

Table 1 compares the trap density obtained from the noise measurements for different kinds of transistors. As seen, trap densities vary by several orders of magnitude depending on the transistor type. Si high-k MOSFETs are characterized by approximately two orders of magnitude higher trap density (and therefore higher noise) than their SiO_2 counterparts. Other type of transistors, like Si TFTs and SiC MOSFETs may have even higher noise and trap density levels, but are still smaller than for the transistors under investigation.

Table 1. Trap densities obtained from noise measurements for different kinds of transistors.

	Trap density N_{tv}, $(cm^{-3}eV)^{-1}$	References
Si MOSFET, SiO_2	$2\times10^{16} - 2\times10^{18}$	[17-21]
Si MOSFET, HfO_2	$10^{18} - 10^{20}$	[22-27]
Amorphous and polycrystalline Si TFTs, SiO_2, SiN_x	$10^{18} - 10^{21}$	[4-6,28,29]
SiC n-MOSFET, SiO_2	$4\times10^{18} - 10^{20}$	[30]
InGaAs, $GdScO_3$	$10^{20} - 10^{22}$	Present work

The interface density of traps, D_{it}, can be also estimated in the framework of the McWhorter model (see [31], for example):

$$\frac{S_I}{I_d^2} = \frac{kTD_{it}}{fL_g W n_s^2 \ln(f_{max} / f_{min})}, \tag{3}$$

where f_{max} and f_{min} are the maximum and minimum frequencies of the $1/f$ noise. Taking for f_{max} and f_{min} the maximum and minimum frequencies of the measurements, we can estimate the interface trap density D_{it} which contributes to noise within the given frequency range. D_{it} relates to N_{tv} as $D_{it}=N_{tv}\Delta t$, where Δt is the oxide depth sensed by the noise measurements. For $f_{max}/f_{min}=10^5$ $\Delta t \approx 10^{-7}$cm, that qualitatively agrees with the estimates for Δt value obtained by other methods [32,33]. Figure 7 compares the interface trap density D_{it} obtained from the noise and conductance measurements. As seen, both methods yield quite high trap density, which increases above threshold with the gate voltage.

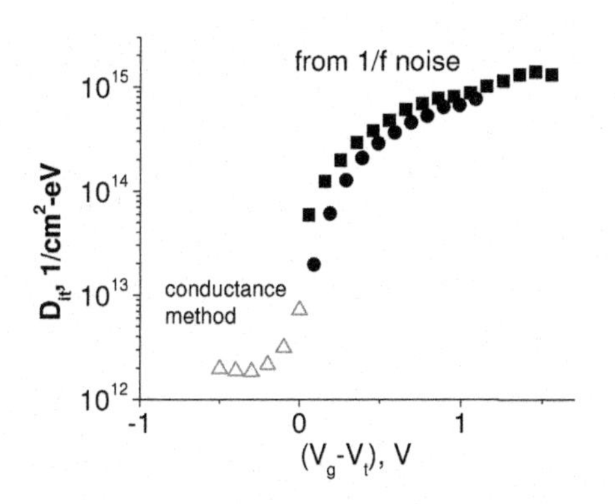

Fig. 7. Comparison of the trap densities obtained from conductance method and from 1/f noise.

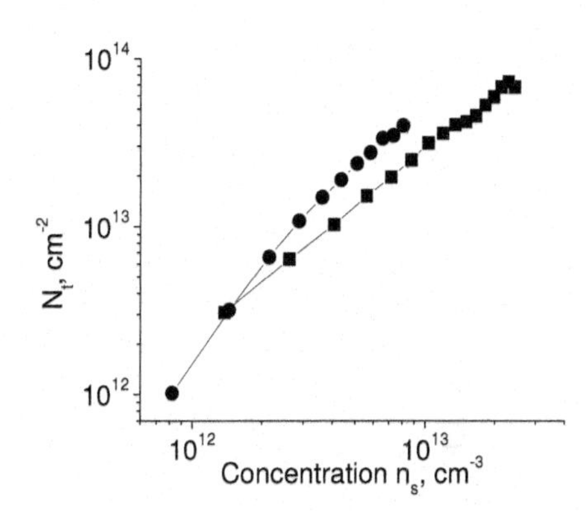

Fig. 8. Trap density $N_t=D_{it}\times 2kT$ as a function of the electron concentration in the channel.

The McWhorter model yields the trap density within an energy ΔE of a few kT around the Fermi level. In Fig. 8, the trap density $N_t=D_{it}\times 2kT$ is plotted versus electron concentration in the channel calculated using eq. (2). As seen, concentration N_t exceeds the channel concentration. Under these conditions the McWhorter model may yield inaccurate values for the trap concentration, therefore, concentrations N_{tv} and D_{it} can be considered in this case only as a figure of merit for the noise amplitude giving the rough estimate of the trap density.

4. Conclusions

Electrical and noise characteristics of n-channel enhancement mode InGaAs MOSFETs with the $GdScO_3$ gate dielectric have been studied at room temperature. Current voltage characteristics demonstrated good linearity at small drain voltage, good saturation at high drain voltage, high on/off ratio, and very low gate leakage current. However, the low frequency noise in these devices was very high indicating a high interface density of traps contributing to noise. A similar problem is known for Si-based high-k MOSFETs, which demonstrate up to two orders of magnitude higher noise level than that MOSFETs with SiO_2 as a gate dielectric. Estimates of the trap density yield values comparable or even exceeding the carrier concentration in the channel. These high values of trap concentration might be over overestimated due to the model limitations but could be considered as figures of merit characterizing noise.

References

1. S. Oktyabrsky, V. Tokranov, M. Yakimov, R. Moore, S. Koveshnikov, W. Tsai, F. Zhu, and J. C. Lee, *Materials Sci. Eng. B* **135**, 272 (2006).
2. R. Kambhampati, S. Koveshnikov, V. Tokranov, M. Yakimov, T. Heeg, M. Warusawithana, D. G. Schlom, W. Tsai, and S. Oktyabrsky, Microelectronic Engineering, (2010) (in press).
3. S. M. Sze, John Wiley & Sons, (1981).
4. J. Rhayem, D. Rigaud, M. Valenza, N. Szydio, and H. Lebrun, *J. Appl. Phys.* **87**, 1985 (2000).
5. L. Pichon, A. Boukhenoufa, C. Cordier, and B. Cretu, *J. Appl. Phys.* **100**, 054504 (2006).
6. M. Rahal, M. Lee, and A. P. Burdett, *IEEE Trans.* ED-**49**, 319 (2002).
7. D. Rigaud, M. Valenza, and J. Rhayem, *IEE Proc. – Circuits, Devices, Syst.* **149**, 75 (2002).
8. S. L. Rumyantsev, M. S. Shur, M. E. Levinshtein, P. A. Ivanov, J. W. Palmour, M. K. Das, and B. A. Hull, *J. Appl. Phys.* **104**, 094505 (2008).
9. A. L. McWhorter, Proc. of the Conf. on the Phys. Semicond. Surfaces, 1956, Philadelphia, pp. 207-29.
10. S. Christensson, I. Lundstrom, and C. Svensson, *Solid State El.* **11**, 797 (1968).
11. K. Lee, M. Shur, T. A. Fjeldly, and T. Ytterdal, Prentice-Hall, Englewood Cliffs, NJ, (1990).
12. M. E. Levinshtein, S. L. Rumyantsev, R. Tauk, S. Boubanga, N. Dyakonova, W. Knap, A. Shchepetov, S. Bollaert, Y. Rollens, and M. S. Shur, *J. Appl. Phys.* 102, 064506 (2007).
13. K. Fobelets, S. L. Rumyantsev, T. Hackbarth, and M. S. Shur, *Solid-State El.* 53, 626 (2009).
14. R. P. Jindal and A. van der Ziel: Solid-State Electron. 21 (1978) 901.
15. G. Reimbold, *IEEE Trans.* **ED-31**, 9 (1984).
16. G. Ghibaudo, O. Roux, Ch. Nguyen-Duc, F. Balestra, and J. Brini, *Phys. Stat. Sol (a)*, **124**, 571 (1991).
17. K. K. Hung, P. K. Ko, C. Hu, and Y. C. Cheng, *IEEE Trans.* **ED-37**, 3 (1990).
18. R. Jayaraman and C. G. Sodini, *IEEE Trans.* **ED-36**, 9 (1989).
19. Y. A. Allogo, M. Marin, M. de Murcia, P. Llinares, and D. Cottin, *Solid St. El.* **46**, 977 (2002).
20. M. von Haartman, Solid-State Electronics. **51**, 771 (2007).
21. M. Fadlallah, G. Ghibaudo, J. Jomaah, M. Zoaeter, and G. Guegan, *Microelectronics Rel.* **42**, 41 (2002).
22. T. H. Morshed, S. P. Devireddy, Zeynep Celik-Butler, A. Shanware, K. Green, J. J. Chambers, M. R. Visokay, and L. Colombo, *Solid-State El.* **52**, 711 (2008).
23. E. Simoen, A. Mercha, L. Pantisano, C. Claey, and E. Young, *IEEE Trans.* **ED-51**, 780 (2004).

24. B. Min, S. P. Devireddy, Zeynep Çelik-Butler, A. Shanware, L. Colombo, K. Green, J. J. Chambers, M. R. Visokay, and A. L. P. Rotondaro, *IEEE Trans.* **ED-53**, 1459 (2006).
25. S. P. Devireddy, B. Min, Zeynep Celik-Butler, H.-H. Tseng, P. J. Tobin, and A. Zlotnicka, *Microelectronics Rel.* **47**, 1228 (2007).
26. P. Srinivasan, F. Crupi, E. Simoen, P. Magnone, C. Pace, D. Misra, and C. Claeys, *Microelectronics Rel.* **47**, 501 (2007).
27. B. Min, S. P. Devireddy, Zeynep Çelik-Butler, F. Wang, A. Zlotnicka, H.-H. Tseng, and P. J. Tobin, *IEEE Trans.* **ED-51**, 79 (2004).
28. A. T. Hatzopoulos, N. Arpatzanis, D. H. Tassis, C. A. Dimitriadis, F. Templier, M. Oudwan, and G. Kamarinos, *Solid-State El.* **51**, 726 (2007).
29. S. L. Rumyantsev, S. H. Jin, M. S. Shur, and Mun-Soo Park, *J. Appl. Phys.* **105**, 124504 (2009).
30. S. L. Rumyantsev, M. S. Shur, M. E. Levinshtein, P. A. Ivanov, J. W. Palmour, M. K. Das, and B. A. Hull, *J. Appl. Phys.* **104**, 094505 (2008).
31. Zeynep Celik-Butler*, Petr Vasina, *Solid-State El.* **43**, 1695 (1999).
32. S. Rumyantsev, C. Young, G. Bersuker, and M. Shur, 20th International Conference on Noise and Fluctuations, ICNF 2009, June 14th-19th, 2009 Pisa, Italy, *AIP Conf. proceedings*, 1129, p. 255.
33. D. Veksler, G. Bersuker, S. Roumiantsev, H. Park, C. Young, K. Y. Lim, W. Taylor, M. Shur, and R. Jammy, *2010 IEEE International Reliability Physics Symposium*, May 2-6, 2010, Anaheim, CA, proceedings.

LOW-POWER BIOMEDICAL SIGNAL MONITORING SYSTEM FOR IMPLANTABLE SENSOR APPLICATIONS

MOHAMMAD RAFIQUL HAIDER

Department of Engineering Science, Sonoma State University, 1801 East Cotati Avenue, Rohnert Park, CA 94928, USA
mhaider407@gmail.com

JEREMY HOLLEMAN

Department of Electrical Engineering and Computer Science, The University of Tennessee, 1508 Middle Drive, Knoxville, TN 37996-2100, USA
jhollema@utk.edu

SALWA MOSTAFA

Department of Electrical Engineering and Computer Science, The University of Tennessee, 1508 Middle Drive, Knoxville, TN 37996-2100, USA
salwa@utk.edu

SYED KAMRUL ISLAM

Department of Electrical Engineering and Computer Science, The University of Tennessee, 1508 Middle Drive, Knoxville, TN 37996-2100, USA
sislam@utk.edu

Implantable biomedical sensors and continuous real time *in vivo* monitoring of various physiological parameters requires low-power sensor electronics and wireless telemetry for transmission of sensor data. In this article, generic blocks required for such systems have been demonstrated with design examples. Ideally neural or electro-chemical sensor signal monitoring units comprise of low noise amplifiers, current or voltage mode analog to digital domain data conversion circuits and wireless telemetry circuits. The low-noise amplifier described here has a novel open loop amplifier scheme used for neural signal recording systems. The design has been implemented using 0.5-μm SOI-BiCMOS process. The fabricated chip can work with 1 V supply and consumes 805 nA. The current mode analog to digital conversion signal processing circuitry takes the current signal as an input and generates a pulse-width modulated data signal. The data signal is then modulated with a high frequency carrier signal to generate FSK data for wireless transmission. The design is fabricated in 0.5-μm standard CMOS process and consumes 1.1 mW of power with 3.5 V supply.

Keywords: Electro-chemical; low-power; neural signal; low-noise; implantable sensor.

1. Introduction

Continuous *in vivo* monitoring of various physiological parameters such as glucose, lactate, pH, CO_2, etc inside the human body as well as neural signal sensing, and remote processing and analysis of sensor data have received prime importance among the researchers in recent years[1]. Self-monitoring or frequent in-clinic monitoring of a patient is logistically difficult and is usually burdened with patient non-compliance. As a result implantable sensors that can monitor the vital information of a patient on a continuous basis are desirable. However an efficient interface between physiological signal from the implantable sensor and external processing tool is needed to successfully transmit the collected information outside of the human body for further diagnosis.

To utilize the benefits of implantable biosensor efficiently, these systems need to occupy very small area and should be capable of operating at very low-power and preferably use non-invasive method for power and data transmission. Johannessen[2] *et al.* demonstrated a system built in 0.6 μm CMOS process that consumed 12.1 mW, and measures 5.5×1.6 cm^2 in area. Mohseni[3] *et al.* reports a multichannel wireless frequency modulated system with total chip area of 4.84 mm^2 and 2.2 mW of power consumption with a 3 V power supply. Most of these approaches are complex and require large chip area. The large power requirements of these designs make them vulnerable to reduce battery lifetime of an implanted sensor. One solution to supply power to an implanted system is inductive powering. In order to facilitate the inductive power-link option to the implanted unit each unit has to exhibit low-voltage and low-power operation. Conventional complex circuit blocks such as operational transconductance amplifier (OTA), mixer, frequency synthesizer, etc are power hungry and require a lot of expensive silicon chip area. In addition, conventional transmitter architectures for cellular communication are not suitable for short-range wireless communication required for transdermal or subcutaneous sensor applications. Therefore an innovative design approach is needed to meet the design requirements of area, power and cost.

In this paper, two integrated circuit systems for monitoring of electro-chemical sensor signal and neural signal from implanted sensors are described`. The neural signal monitoring unit describes the design and implementation of a low-power and low-noise amplifier configuration for neural signal recording systems. The open loop configuration of the neural signal amplifier demonstrates sub-μW range power consumption and excellent input referred noise performance. The design has been implemented using a 0.5-μm SOI-BiCMOS process and the chip occupies an area of 0.033 mm^2. Relatively simple and area-efficient CMOS inverter architecture has been adopted to build the signal processing block and a carrier generator block of the electro-chemical sensor system described here to reduce the voltage and power requirement. The system has been fabricated using 0.5-μm standard CMOS process and the fabricated chip occupies an area of only 0.046 mm^2.

2. Biomedical Sensors and Signal Monitoring

Various types of sensors are usually used for biomedical applications. Neural signals from brain implants are typically voltage spikes whereas metabolic monitoring signals from electro-chemical sensors are amperometric. In biomedical applications these can include things such as muscle displacement, blood pressure, core body temperature, blood flow, cerebrospinal fluid pressure, bone growth, etc. For metabolic monitoring the well known sensor type is electro-chemical sensors which generate current signal in response to a particular analyte's presence or concentration in intravenous fluid of human body. Usually two or three electrode based electro-chemical sensors are placed inside the human body and biased with a constant potential by using a potentiostat. As a result the generated current signal from the electrodes is a perfect representation of concentration variation of the particular chemical compound such as glucose, lactate, pH, CO_2, etc in the body fluid.

Detection of neural activities of human brain or tissue is usually performed using micro- or nano-electrodes array. The signal path in a neural recording system must typically start with an amplifier in order to boost the signal levels and buffer the high source impedance. Because of extremely small signal amplitudes, amplifier noise must be minimized in order to avoid further degradation of the signal. Additionally, the high impedance of neural electrodes necessitates a high impedance input. For a fixed bandwidth, an amplifier's input-referred noise scales inversely with the square of its current consumption. In order to achieve acceptable noise levels, the front-end amplifier often consumes a substantial fraction of the overall system power[4]. Recently there has been a great deal of research into the design of low-power amplifiers for neural recording[5,6,7].

In the following sections circuit design and development of each type of unit in an implantable system is discussed.

3. Biomedical Sensor Signal Monitoring System Architecture

The system typically consists of three core blocks, a signal conditioning front end, a signal processing block and a telemetry circuit. The signal conditioning block, which often consists primarily of a low-noise amplifier, is needed because the small signal amplitudes and high source impedances typical of bio-electronic interfaces must be buffered and amplified to achieve reasonable precision. The signal processing block converts the signal into a format suitable for telemetry and may also perform some local computation. In an implanted device, a telemetry system is needed to deliver the data to a base station, data collection unit, or actuation system. In this section we will describe implementations of each of the three core blocks.

3.1. *Low noise amplifier*

The signal path in a biomedical monitoring system must typically start with an amplifier in order to boost the signal levels and buffer the high electrode impedance. Because of extremely small signal amplitudes (e.g. 100-200 μV for extracellular neural spikes), amplifier noise must be minimized in order to avoid further degradation of the signal. Additionally, the high impedance of neural electrodes necessitates a high impedance input.

For a fixed bandwidth, an amplifier's input-referred noise scales inversely with the square of its current consumption. In order to achieve acceptable noise levels, the front-end amplifier often consumes a substantial fraction of the overall system power[4]. Due to the difficulty of delivering power to an implanted system, it is desirable that the amplifiers be designed to operate with minimal power consumption. In this section we discuss a novel topology for improving the noise-power efficiency of neural amplifiers.

In order to minimize input-referred noise while meeting a fixed bandwidth and power consumption specification, an amplifier should be designed with two considerations in mind. First, as much of the total bias current as possible should be used in the amplifier's first stage, since the noise of the first stage will dominate overall noise. Secondly, the first stage should be designed to maximize transconductance per unit current. Current noise at the output of the first stage will then be minimized when referred back to the input. The open-loop amplifier described here carries these two considerations to their natural limit.

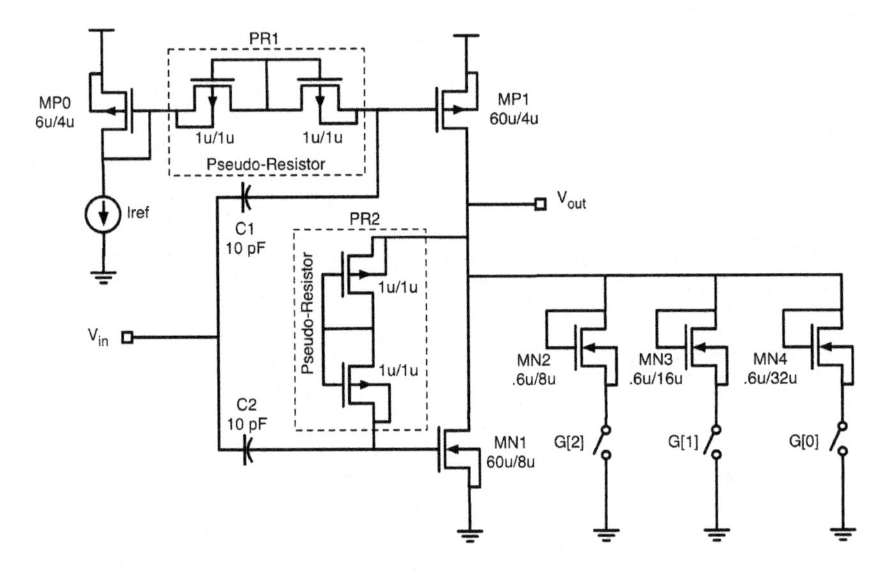

Fig. 1: Open-loop amplifier schematic.

A single-ended, open-loop amplifier[8], designed for recording action potentials is shown in Fig. 1. MOS-bipolar pseudo-resistors (PR)[5] are used to implement the AC coupling necessary to reject large DC offsets due to contact potentials.

Two strategies are utilized here to minimize the input-referred noise for a given bias current. The first is to limit the number of current branches. There is only one branch operating at full current. The reference current is ten times smaller than the amplifier bias current, so it does not contribute significantly to the total power consumption. The same RC network used to AC couple the PMOS input presents a low-pass filter to the reference transistor MP0, so noise from the current reference is not added to the signal, permitting the use of a relatively noisy low-power bias generator.

The second strategy is to drive the gates of both MP1 and MN1. A conventional common-source amplifier has a current-source load which adds noise to the signal, but performs no amplification. Because the input must be AC-coupled, it is possible to decouple the DC levels of the gates of transistors MP1 and MN1 while keeping them connected in the frequency band of interest. The transconductance of the amplifier is effectively doubled, while output noise remains constant, reducing the input-referred noise voltage spectral density by a factor of two. Because the bandwidth is determined by the load capacitor and is set based on the application requirements, the input-referred RMS noise voltage is also reduced by a factor of two.

The aspect ratios of MP1 and MN1 were chosen to place both transistors in the weak inversion regime in order to maximize g_m/I_D. The lengths of the transistors MP1 and MN1 were chosen to be large to obtain sufficient gain from a single stage and to reduce 1/f noise. The bias current is generated from an on-chip bias circuit[9] and multiplied by a 3-bit digitally-controlled current mirror. The bias current in the amplifier can be varied from 110 nA to 770 nA. The biasing circuitry also includes a bank of digitally-enabled diode-connected transistors M2-M4, allowing the user to control the gain through the gain-control word G[0:2], in order to mitigate the risk of the gain variation due to process variation.

3.2. *Signal processing and telemetry*

The signal processing block discussed here processes current from the sensor input and produces a digital binary output signal whose frequency is proportional to the sensor current. The telemetry circuit takes the output from the signal processing block as input and modulates it to FSK signal using two different frequencies for binary high and low inputs. Fig. 2 represents the block diagram of the system and shows how the biosensor is incorporated with the microelectronic circuit blocks.

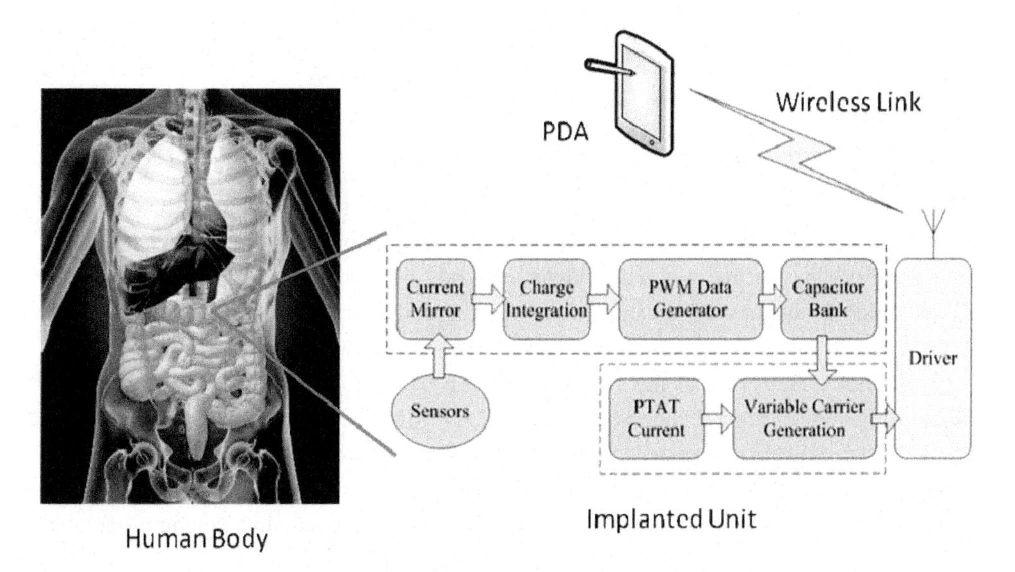

Fig. 2: Implantable sensor and signal processing.

3.2.1. *Current to frequency converter block*

Fig. 3 shows the schematic diagram of the signal processing block. Sensor current is fed to the first stage which is a current mirror to isolate the sensor output current from the rest of the circuitry. The output from the current mirror charges the capacitor. As the capacitor charges, the voltage at the input terminal of the Schmitt trigger decreases and when it reaches the lower threshold voltage, the Schmitt trigger triggers and produces a positive DC voltage. The output of the Schmitt trigger through the inverter turns on the PMOS transistor creating a discharge route for the capacitor. As the capacitor discharges the Schmitt trigger input node voltage increases and when it reached the upper threshold value the Schmitt trigger's output goes negative, turning off the PMOS. As the discharge path is now open, the capacitor starts charging again and another cycle begins. The period of this cycle depends on the charging time of the capacitor, which is proportional to the charging current, i.e. the input current from the sensor.

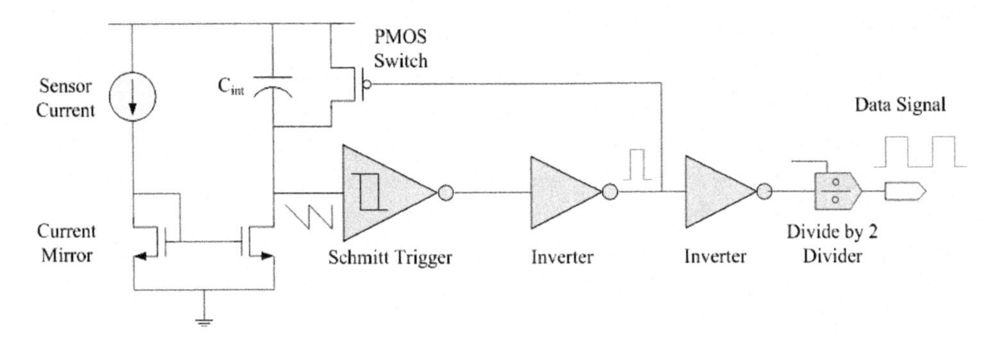

Fig. 3: Signal processing block.

3.2.2. *FSK transmitter block*

The detailed block diagram of the FSK transmitter Block is shown in Fig. 4. It consists of a Schmitt trigger, three inverters and a switched-capacitor structure. Based upon the functionality, the entire modulator block can be subdivided into "Carrier Generator Block" and "Frequency Modulator Block". The detail descriptions of these two blocks are as follows.

A similar architecture of the signal processing block is used in the carrier generator block to generate high frequency (1 MHz-1.67 MHz) carrier signal where instead of the sensor current a proportional-to-absolute-temperature (PTAT) current is used to generate the fixed frequency signal. Finally the data signal and the carrier signal are combined using a switch-capacitor structure to achieve frequency-shift-keying modulated signal.

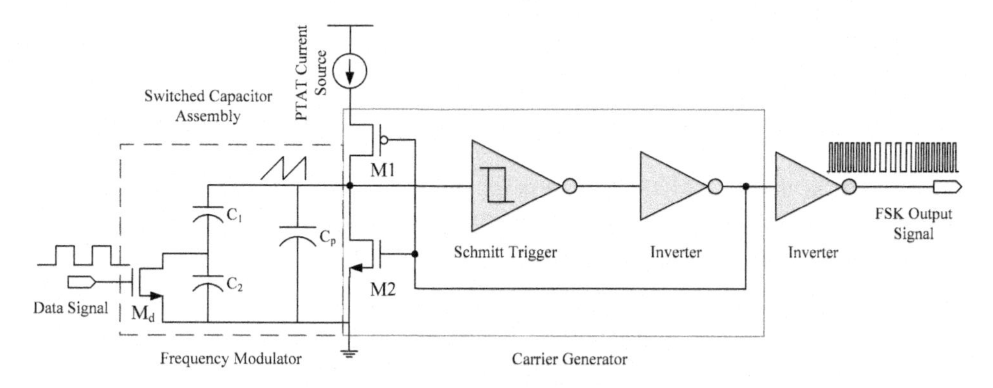

Fig. 4: FSK transmitter block.

PMOS (M1) and NMOS (M2) transistors form an inverter at the input of the Schmitt trigger. When the output of the inverter is '1', the constant current from an external source charges the capacitor assembly connected to the inverter output to supply voltage. The charging period depends on the equivalent capacitor value and the equivalent resistance of the PMOS. When the inverter output voltage reaches the upper threshold value of the Schmitt trigger, it produces a negative pulse or '0', which after passing through another inverter feeds a logic '1' input to the inverter. Inverter output goes low and the capacitor assembly starts discharging through the NMOS (M2). Here also the discharging rate depends on the equivalent capacitor value and the equivalent resistance of the NMOS. When the output voltage reaches the lower threshold value of the Schmitt trigger, it produces a positive voltage output, which presents a "low" or '0' input to the M1- M2 inverter. The inverter input then goes high again and the entire cycle is repeated. Since the capacitor value remains unchanged during this cycle, the duty cycle of the generated signal is determined by the aspect ratio of the NMOS and PMOS transistors. In this design, the NMOS and PMOS transistors are carefully chosen to achieve a 50% duty cycle.

The switched-capacitor assembly consists of three capacitors C_1, C_2 and C_p in a series parallel combination. An NMOS (M_d) transistor is placed in parallel with the capacitor C_2. The gate of this MOSFET is driven by the digital output signal from the signal processing block. When the signal is 'low', NMOS acts like an open circuit and circuit sees the capacitor value of C_2. The equivalent capacitance value in this condition is expressed by

$$C_{eq,low} = C_p + \frac{C_1 \cdot C_2}{C_1 + C_2} \tag{1}$$

So, the time period required for charging and discharging is

$$\tau_{low,charging} = R_{PMOS} \times C_{eq,low} \tag{2}$$

$$\tau_{low,discharging} = R_{NPMOS} \times C_{eq,low} \tag{3}$$

The equivalent frequency may be calculated by

$$f = \frac{1}{\tau_{low,charging} + \tau_{low,discharging}} \tag{4}$$

When signal is 'high', NMOS (M_d) turns on and short circuits the C_2 capacitor. The equivalent capacitance value in this condition is expressed by

$$C_{eq,high} = C_p + C_1 \tag{5}$$

Similarly for this condition, frequency can be calculated as

$$f = \frac{1}{\tau_{high,charging} + \tau_{high,discharging}} \tag{6}$$

From Eq. (1), (2) and (3), it is obvious that whenever the data signal is 'low', the equivalent capacitance value is lower due to the series combination of C_1 and C_2. Thus the charging and the discharging time periods are smaller and consequently the output frequency is higher. On the other hand, whenever the data signal is 'high', the equivalent capacitance value is higher and ultimately the output frequency is lower. Thus the modulator block generates carrier signal of two different frequencies depending on the input from the signal processing block. Hence an FSK modulated signal is produced as the final output from the system which can be fed to an on-chip antenna for transmission.

4. Measurement Results

4.1. *Amplifier results*

The neural amplifier was fabricated in a 0.5-μm SOI-BiCMOS process, employing CMOS devices exclusively. It occupies .033 mm^2 and the current reference occupies an additional .013 mm^2 of die area. The entire circuit can operate from a supply between 1 V and 5 V, while the measurements presented here were taken with a 1.0 V supply.

4.1.1. *Neural amplifier measurements*

Fig. 5(b) shows the frequency response over the entire range of gain settings. The current reference is configured to provide the maximum bias current, yielding I_{DS} = 770 nA for MP1 and MN1. At the highest gain setting, the amplifier exhibits a gain of 44 dB and bandwidth of 1.9 KHz. The intermediate gain setting provides a gain of 38 dB and a 3 dB frequency of 3.6 KHz. With the lowest gain setting, the gain is 36 dB, and the 3 dB frequency is extended to 4.7 KHz. The remainder of this section will focus primarily on the low-gain setting, because it provides sufficient bandwidth to record action potentials. However, it is possible to extend the bandwidth at higher gain settings by increasing the bias current, either by overriding the internal bias generator, or with a modified design.

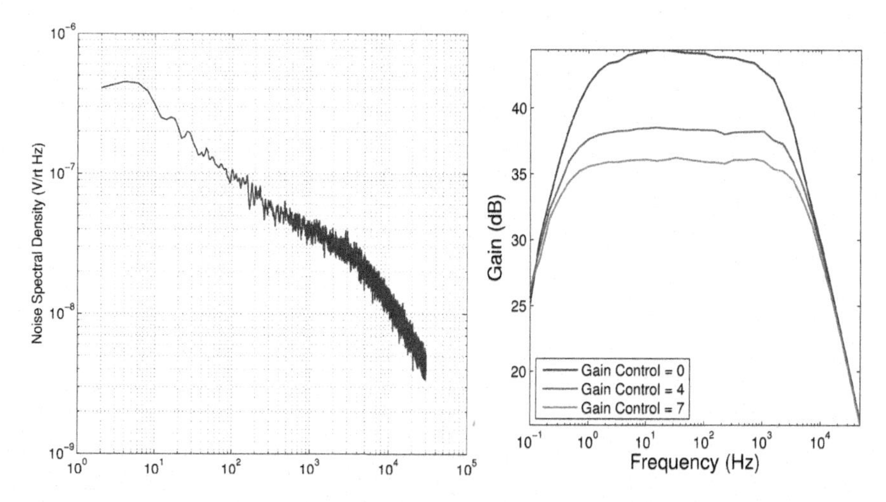

Fig. 5: (a) Input-referred noise spectrum of the open-loop amplifier. (b) Frequency response of the open-loop amplifier gain for three gain control settings.

The input-referred noise spectrum of the amplifier is shown in Fig. 5(a). The total RMS noise at the input is 3.5 μV. Feedback amplifiers achieve high linearity because their gain is determined by ratios of passive components. For open-loop amplifiers, nonlinearity of the transconductance and of the output impedance is manifested in a nonlinear input-output function. The open-loop amplifier's total harmonic distortion (THD) with a 10 mV input is 18.12%, but for a 1mVpp input, THD is lower, at 7.06% and 6.63% for the low and high gain settings, respectively.

In applications where a quiet power supply cannot be guaranteed, power-supply rejection ratio (PSRR) must be examined. In the proposed amplifier, both MP1 and MN1 have their sources connected to a power supply and their gates capacitively connected to the input. Thus, the positive and negative supplies directly modulate the P- and N-type transconductors, respectively. Therefore one would expect that the gain from the positive power supply to the output will be approximately half the gain from input to output, resulting in a minimal PSRR of 6 dB. Measurements show a PSRR of 5.5 dB.

Table 1 compares the performance of this amplifier to other published bio-signal amplifiers. The noise efficiency factor (NEF)[10] normalizes amplifier noise based on amplifier bandwidth and current consumption to allow comparisons between amplifiers optimized for different parameters. The proposed amplifier demonstrates the lowest NEF of any amplifier reported to date. Including the bias circuitry, the entire amplifier chip dissipates less than 1 μW.

4.1.2. *Amplifier discussion*

The choice between a single-ended open-loop amplifier and a differential closed-loop amplifier depends on system-level considerations. The primary drawbacks of the open-loop LNA are gain inaccuracy, nonlinearity, and poor supply rejection. Because absolute amplitude is not typically a salient feature of neural recordings (due partially to other sources of amplitude uncertainty), the decision can be made based on linearity, supply rejection and power consumption considerations. Power supply rejection requirements will vary greatly depending on specific system configurations. In the case of high channel-count systems, it may be justifiable to focus the supply rejection burden on a single regulator that can be shared across many channels.

Table 1: Amplifier Comparison.

	Gain (dB)	I_{DD}	NEF	$V_{ni,rms}$	THD (@ Input)	PSRR (dB)	Bandwidth
Harrison[4]	39.5	16μA	4.0	2.2μV	1% @ 16.7mVpp	> 85	.025Hz-7.2KHz
Denison[6]	45.5	1.2A	4.9	.93μV	—	—	.5Hz-250Hz
Wu[7]	40.2	330nA	3.8	.94μV	.053% @ 5mVpp	62	3mHz-245Hz
T his work							
Open-loop	36.1	805nA	1.8	3.6μV	7.1% @ 1mVpp	5.5	.3Hz-4.7KHz

The very small amplitude of many biological signals, including neural spikes, relaxes the linearity requirements of an amplifier. Fig. 6(a) uses a measured voltage transfer function to simulate the effect of non-linear amplification on a single spike. With a typical 150 μV amplitude, there is no noticeable difference between the two, indicating that the nonlinearity of the open-loop amplifier will not contribute to significant error for such small signals.

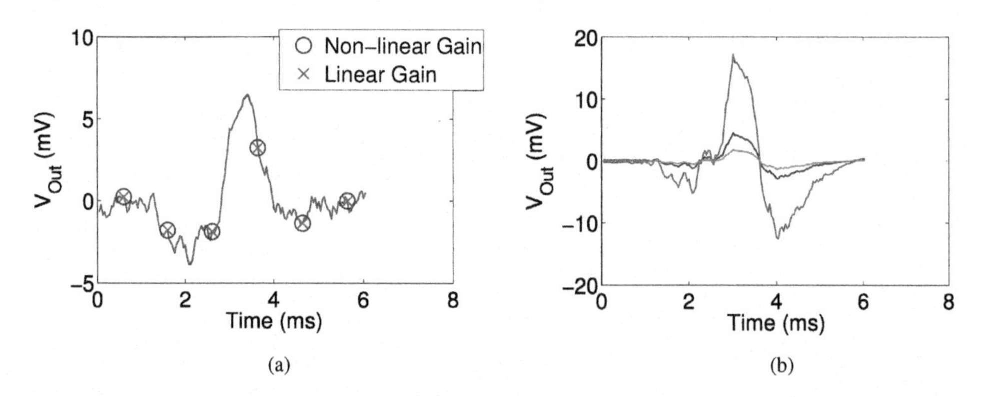

Fig. 6: (a) Simulated comparison of a single spike amplified by linear and non-linear gain. (b) Effect of non-linear gain in the presence of large interfering signals.

However, extracellular neural probes will also detect local field potentials (LFPs), which occur in the frequency band below 1 KHz and can have amplitudes as high as 5 mV. Additionally, 50Hz/60Hz interference from wall power may corrupt the signals. If LFPs are in the amplifier's passband, they will shift any action potentials to a different point in the amplifier's input/output curve, effectively causing the relevant part of the input signal (the action potentials) to experience time-varying gain. In Fig. 6(b), three spikes are added to a 5 mV sinusoid before being applied to the amplifier nonlinearity, to simulate the effect of a small spike being recorded in the presence of a large LFP interferer. In this case, the amplifier's nonlinearity does introduce significant errors, essentially subjecting the desired action potential signal to a time-varying gain.

If large interfering signals are not present then linearity may be sacrificed for power consumption. In the presence of interferers, filtering should be added at the input of the amplifier. If that is not possible, the superior linearity of a closed-loop topology might be preferable. The complementary drive technique demonstrated in the open-loop amplifier here can also be incorporated into a closed-loop amplifier[10].

The relatively low frequencies typical of biomedical signals and the large transistor sizes dictated by flicker noise considerations suggest that older, inexpensive CMOS processes are appropriate. When bio-signal amplifiers are to be included in a highly integrated system with large digital components, it may be desirable to use more advanced CMOS processes. Because of the extremely high incremental resistance of the DC biasing pseudo-resistors, it is important to avoid any significant gate current at the amplifier input. Fortunately, most modern processes offer a high-voltage transistor with thicker oxide for I/O, which reduces gate current to negligible levels. With judicious use of I/O transistors, it should be possible to port most bio-amplifier designs to modern CMOS processes.

4.2. *Signal processing and telemetry results*

The signal processing and telemetry circuit for electro-chemical sensor has been designed and fabricated using 0.5-μm standard CMOS process. The fabricated chip occupies an area of only 0.046 mm^2.

The signal processing circuit produces reliable data (960 Hz to 150 KHz) for the sensor current in the range of 20 nA to 5 μA with only 1.3 V power supply and consumes 626 μW of power. The carrier generator block operates reliably at 3.5 V supply voltage. The power consumption of the entire system including signal processing block and the carrier generator block is 1.1 mW.

The fabricated chip was tested by applying constant dc current as sensor current input from a Keithley 2400 source meter. A constant current supply of 10 μA was fed to the modulator input of the carrier signal generating block from another source meter. Static power dissipation was measured by putting an ammeter in series with the V_{DD} power supply and multiplying the dc voltage and the dc current supplied by the voltage source.

The output from the signal processing block for sensor input current of 1 μA and the corresponding FSK output is shown in Fig. 7. At high frequency the amplitude of the modulated signal is reduced because of the parasitic capacitance associated with the process technology and the packaging. At high frequency there is more parasitic loss and the circuit output is not strong enough to drive the external capacitance associated with the chip package pins and the breadboard connections. Table 2 shows the data frequency variation with different sensor current levels. Fig. 8 is a graphical representation of sensor input current versus the data signal frequency and shows the linearity of the system with R-squared value equal to 0.9984. The duty cycle of the data signal is around 50%.

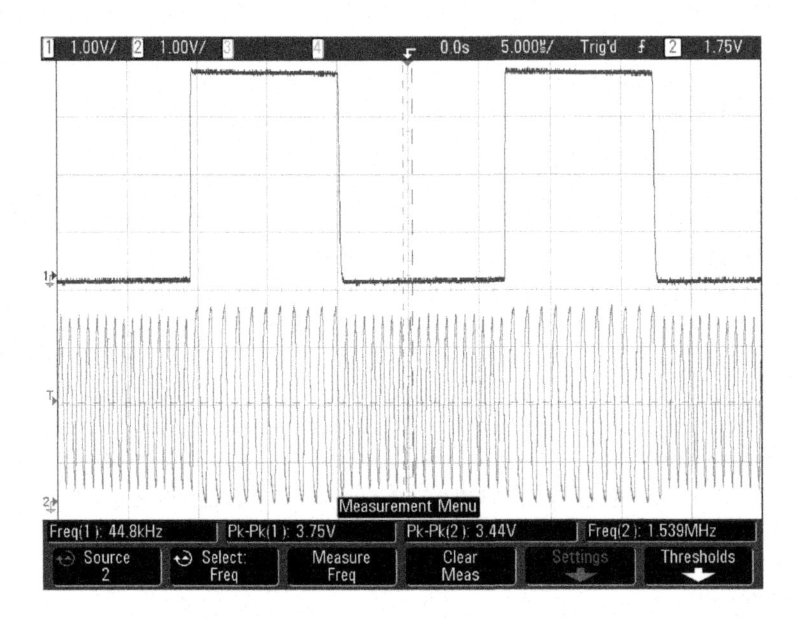

Fig. 7: Data signal and FSK output for sensor current of 1 μA.

Table 2: Data signal frequency with respect to varied sensor current.

Sensor Current	Data frequency	Duty cycle
μA	KHz	%
0.02	0.960	35.6
0.04	1.85	38
0.05	1.53	39.3
0.1	4.76	39.5
0.2	9.22	40.1
0.5	22.4	43.8
0.8	37.0	46.7
1	44.9	47.4
1.2	54.6	48.1
1.5	65.4	48.4
1.8	79.2	49.1
2	89	48.6
2.5	109	49.1
3	129	49
3.5	149	49.7
4	167	49.3
4.5	184	49.8
5	204	49.8

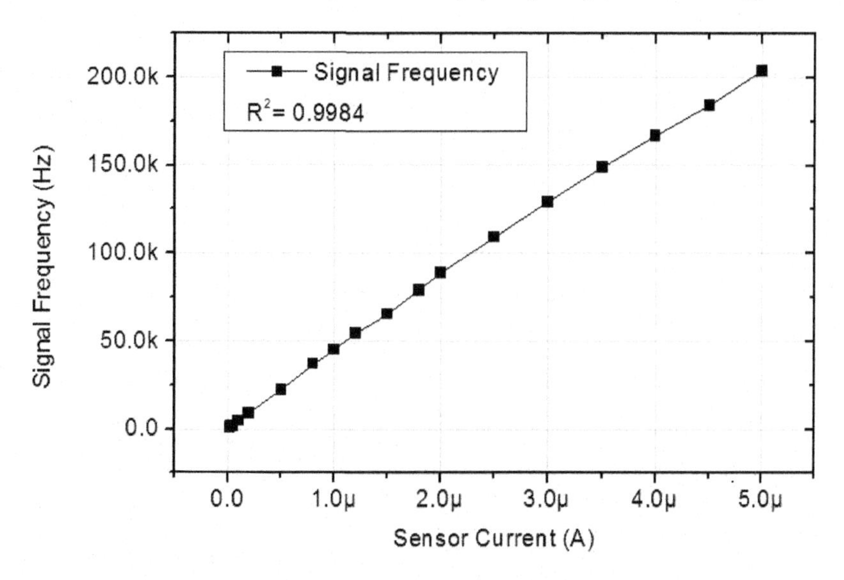

Fig. 8: Data frequency variation with different sensor current levels.

The system has been designed using strong inversion MOSFETs with careful consideration of MOSFET dimension to achieve low-voltage and low-power consumption. Strong inversion operation offers noise immunity, high linearity and high speed operation. Relatively low frequency and low data rate application of biomedical implants make it extremely suitable to use weak inversion MOSFET to further scale down the voltage and power requirement. However weak inversion operation suffers from nonlinearity and current mismatch. Therefore future goal will be to implement the system using weak inversion devices with better optimization of power and noise performance. Modern higher-end CMOS process will also be a good candidate to implement the design with ultra-low-power consumption and reduced parasitic effects.

5. Conclusion

We have presented results from two chips demonstrating low-power circuits to implement each of the three core blocks common to most bio-signal sensors. In each case the proposed systems have demonstrated the potential for simple circuit topologies to yield reductions in power consumption.

The proposed open-loop neural signal amplifier provides excellent noise and power consumption in exchange for diminished linearity and supply rejection. The circuit techniques are applicable to many biomedical monitoring applications, as well as other power-constrained sensing problems. Additionally we presented a discussion of the linearity requirements for neural signal amplifiers.

A low-power low-voltage sensor read out circuit with FSK telemetry has been designed and fabricated. The circuit operates at a voltage lower than that required for the process, reducing power consumption. Circuit is built on a very simple topology, which is less power hungry than the complex designs. Test results demonstrate the excellent low-power operation with high degree of linearity performance of the designed circuit for future application in implantable sensors.

References

1. M. Zhang, M. R. Haider, M. A. Huque, M. A. Adeeb, S. Rahman, and S. K. Islam, "A low-power sensor signal processing circuit for implantable biosensor applications," *Smart Materials and Structures*, vol. 16, pp. 525-530, Mar. 2007.
2. E. A. Johannessen, L. Wang, L. Cui, T. B. Tang, M. Ahmadian, A. Astaras, S. Reid, P. Yam, A. Murray, B. Flynn, S. Beaumont, D. Cumming, and J. Cooper, "Implementation of multichannel sensors for remote biomedical measurements in a microsystem formats," *IEEE Trans. on Biomedical Engineering*, vol. 51, no. 3, pp. 525-535, Mar. 2004.
3. P. Mohseni, K. Najafi, S. J. Eliades, and X. Wang, "Wireless multichannel biopotential recording using an integrated FM telemetry circuit," *IEEE Trans. on Neural Systems and Rehabilitation Engineering*, vol. 13, no. 3, pp. 263-271, Sep. 2005.
4. R. Harrison, P. Watkins, R. Kier, R. Lovejoy, D. Black, B. Greger, and F. Solzbacher, "A low-power integrated circuit for a wireless 100-electrode neural recording system," *IEEE Journal of Solid-State Circuits*, vol. 42, no. 1, pp. 123-133, 2007.
5. R. Harrison and C. Charles, "A low-power low-noise CMOS amplifier for neural recording applications," *IEEE Journal of Solid-State Circuits*, vol. 38, no. 6, pp. 958-965, 2003.
6. T. Denison, K. Consoer, A. Kelly, A. Hachenburg, and W. Santa, "A 2.2μW 94nV/√Hz chopper-stabilized instrumentation amplifier for EEG detection in chronic implants," *IEEE International Solid-State Circuits Conference Digest of Technical Papers*, pp. 162-163, 2007.
7. H. Wu and Y. Xu, "A 1V 2.3μW biomedical signal acquisition IC," *IEEE International Solid-State Circuits Conference Digest of Technical Papers*, pp. 119-128, 2006.
8. J. Holleman and B. Otis, "A sub-microwatt low-noise amplifier for neural recording," 29th Annual International Conference of the IEEE Engineering in Medicine and Biology Society, 2007. (EMBS 2007), pp. 3930-3933, 2007.
9. E. Camacho-Galeano, C. Galup-Montoro, and M. Schneider, "A 2-nW 1.1-V self-biased current reference in CMOS technology," *IEEE Trans. on Circuits and Systems II: Express Briefs*, vol. 52, no. 2, pp. 61-65, 2005.
10. S. Rai, J. Holleman, J. Pandey, F. Zhang, and B. Otis, "A 500μW neural tag with 2μVrms AFE and frequency multiplying MICS/ISM FSK transmitter," *IEEE International Solid-State Circuits Conference Digest of Technical Papers*, pp. 212-213, February 2009.

NANOMATERIALS AND NANODEVICES

III-V COMPOUND SEMICONDUCTOR NANOWIRES FOR OPTOELECTRONIC DEVICE APPLICATIONS

Q. GAO, H. J. JOYCE, S. PAIMAN, J. H. KANG, H. H. TAN

Department of Electronic Materials Engineering, Research School of Physics and Engineering, The Australian National University, Canberra, ACT 0200, Australia
gao109@physics.anu.edu.au

Y. KIM

Department of Physics, Dong-A University, Busan 604-714, Korea

L. M. SMITH, H. E. JACKSON

Department of Physics, University of Cincinnati, Cincinnati, Ohio 45221-0011, USA

J. M. YARRISON-RICE

Department of Physics, Miami University, Oxford, Ohio 45056, USA

J. ZOU

School of Engineering and Centre for Microscopy and Microanalysis, The University of Queensland, Brisbane, QLD 4072, Australia

C. JAGADISH

Department of Electronic Materials Engineering, Research School of Physics and Engineering, The Australian National University, Canberra, ACT 0200, Australia
cxj109@physics.anu.edu.au

GaAs and InP based III-V compound semiconductor nanowires were grown epitaxially on GaAs (or Si) (111)B and InP (111)B substrates, respectively, by metalorganic chemical vapor deposition using Au nanoparticles as catalyst. In this paper, we will give an overview of nanowire research activities in our group. In particular, the effects of growth parameters on the crystal structure and optical properties of various nanowires were studied in detail. We have successfully obtained defect-free GaAs nanowires with nearly intrinsic exciton lifetime and vertical straight nanowires on Si (111)B substrates. The crystal structure of InP nanowires, i.e., WZ or ZB, can also be engineered by carefully controlling the V/III ratio and catalyst size.

Keywords: Nanowires; III-V compound semiconductors; metalorganic chemical vapor deposition; carrier dynamics; photoluminescence.

1. Introduction

Semiconductor nanowires are intensively being investigated due to their unique growth mechanisms and novel properties. The 1-dimensional anisotropic growth via vapor-liquid-solid (VLS) mechanism was first developed by Wagner and Ellis more than 40 years ago when they grew um-sized Si whiskers on Si wafers using Au particles as the catalyst.[1] Recently, this technique has been extended to nm-sized nanowire growth on a variety of semiconductor materials including III-V, II-VI compound and group IV elemental semiconductors. With a reduction in size, unique electrical, mechanical, chemical and optical properties have been reported in nanowires, which are largely believed to be the result of small contact area between nanowires and substrate, large surface area over volume and quantum confinement effects. III-V compound semiconductor nanowires are particularly promising for optoelectronic applications, due to the direct band gap and high carrier mobility of these materials. In particular, it is of great interest to epitaxially grow III-V semiconductor nanowires on Si platform to integrate the superior optoelectronic and microelectronic properties from III-V semiconductors and Si, respectively. Various nanowire-based devices, including nanowire lasers,[2-3] photodetectors[4] and biosensors[5] have already been demonstrated in the past decade.

The development of III–V nanowire based devices depends strongly on the ability to grow nanowires with high level of control over material qualities such as morphology (shape, diameter and facets etc), crystal structure and composition. The challenge is to produce nanowires free of crystallographic defects, with uniform diameters, with desired crystal structure and with high purity. However, twin defects, stacking faults, tapered nanowire shape and mixed crystal phases (i.e., zincblende and wurtzite mixture) are the most common problems in III-V semiconductor nanowires grown via VLS mechanism by metalorganic chemical vapor deposition (MOCVD) or molecular beam epitaxy.[6-8] In addition, it adds more complexity to integrate III-V semiconductor nanowires on Si due to lattice mismatch, different thermal expansion coefficients, and the interruption from native oxide layers.

In this paper, we will review several techniques we have developed to achieve epitaxial defect-free GaAs nanowires with high purity and minimal tapering on GaAs substrates, and vertical straight GaAs nanowires on Si using MOCVD. In addition, we demonstrated that the crystal structure of InP nanowires can be engineered into either zincblende (ZB) or wurtzite (WZ) phase or mixed phases. A number of growth conditions, including growth temperature, V/III ratio and growth rate have been tailored for desired nanowire growth. Some unexpected benefits have been discovered in the nanowire growth compared to the conventional 2D planar epitaxy. We have also studied optical properties of various nanowires.

2. Experimental

In this study, GaAs and InP nanowires were grown on GaAs (or Si) (111)B and InP (111)B substrates, respectively, by using a horizontal flow MOCVD reactor operating at a pressure of 100 mbar with ultra high purity hydrogen as the carrier gas. The substrates were immersed in poly-L-lysine (PLL) solution for 60 s and rinsed with deionized water before being applied a droplet of colloidal solution of Au particles from 20 to 50 nm in diameter. After rinsing off the excessive Au colloidal solution with DI water, the wafers were blown dry using N_2 gas and transferred into the MOCVD reactor chamber, and annealed under AsH_3 or PH_3 at 600 °C for 10 min, to desorb contaminants on the surface. Then, the reactor was cooled down to desired temperature to carry out the nanowire growth for 20~30 mins. To grow GaAs nanowires on Si substrates, prior to the PLL treatment, Si substrates were etched by a buffered HF solution and then a thin GaAs buffer layer was deposited. Details of this GaAs buffer layer growth can be found in Ref. 9. For the two-temperature procedure, growth initiated with a 1 min "nucleation" step at the nucleation temperature, T_n, of 450 °C. The temperature was then rapidly ramped down to the subsequent growth temperature, T_g, between 350 and 390 °C. To study the V/III ratio effect, gas flow of group III precursors, TMGa or TMIn, was kept constant and only the gas flow of group V precursors, AsH_3 or PH_3, was varied. To study the effect of growth rate, V/III ratio was kept constant during the change of gas flow of group III precursor. In the meantime, growth time was scaled inversely with group III flow. This was to achieve nanowires of comparable height, between 2 and 5 μm, across all samples.

The structural and crystallographic properties of nanowires were characterized by using field-emission scanning electron microscopy (FESEM) (Hitachi S4500) and transmission electron microscopy (TEM) (FEI Tecnai F30). For single nanowire photoluminescence (PL) measurements, nanowires were transferred from the as-grown substrates to a Si substrate by hand-rubbing the two substrate surfaces together. The PL spectra of individual nanowires were obtained at 18 K using slit confocal micro-photoluminescence spectroscopy. A 50×/0.5 NA long working distance microscope objective was used to project the PL image of the nanowire sample onto the entrance slit of a CCD camera or the spectrometer. In order to obtain a strong PL emission from GaAs nanowires, an AlGaAs shell was grown for 20 min at 650 °C to passivate GaAs core and a thin GaAs cap layer was grown for 5 min at 650 °C to avoid oxidation of AlGaAs shell.

3. Results and Discussion

3.1. *GaAs nanowires on GaAs substrates*

Fig. 1 illustrates FESEM images of some typical GaAs nanowires grown on GaAs substrates at various temperatures. Table 1 summarizes the growth temperature and key characteristics of GaAs nanowires grown in this study. Nanowires grown by the single-temperature procedure were generally straight and epitaxially aligned in the vertical [111]B direction when grown at T_g of 410 °C and above (Fig. 1a), but suffer severe tapering. At T_g of 390 °C and below (Fig. 1b), nanowire growth rarely initiated in the vertical [111]B direction, and subsequent kinking was common: the initial and final nanowire orientations exhibit no apparent relationship with the substrate. In contrast, the two-temperature procedure allowed the growth of straight, vertical [111]B-oriented nanowires at T_g as low as 350 °C (Fig. 1c). We determine a minimum T_n of 410 °C, and a minimum T_g of 350 °C are required for this straight epitaxial nanowire growth.[10]

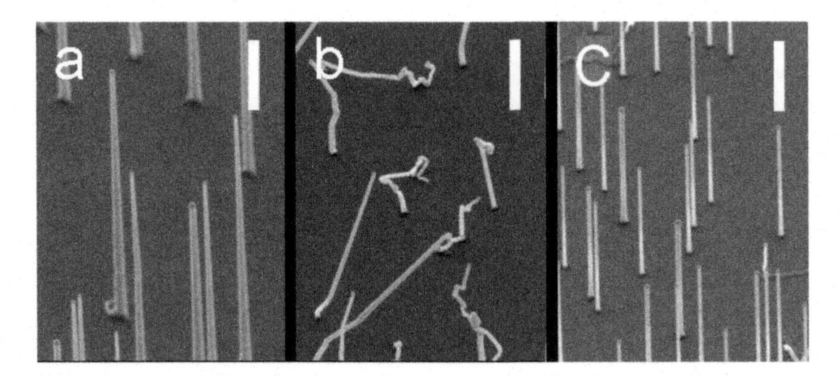

Fig. 1. FESEM images of GaAs nanowires grown by single temperature and two-temperature procedures at various T_g. (a) Single-temperature procedure with T_g of 450 °C. (b) Single temperature procedure with Tg of 390 °C. (c) Two-temperature procedure with T_g of 350 °C (T_n of 450 °C). Samples are tilted at 40°. Scale bar is 2 μm.

Table 1. Summary of nanowire morphology for single- and two-temperature procedures with various growth temperatures (Tg). Tapering is calculated for straight nanowires only.

Procedure	T_n (°C)	T_g (°C)	Straight [111] nanowires (% of total)	Tapering (nm/μm)
1-temperature		450	98	17
1-temperature		390	1	
2-temperature	450	350	88	<2

The temperature-dependent reduction in tapering is explained with reference to the axial and radial growth mechanisms.[11] Reaction species which impinge directly upon the nanoparticle contribute to axial growth. Additionally, Ga adatoms are adsorbed on the substrate and nanowire sidewalls and diffuse along the concentration gradient toward the growing nanoparticle- nanowire interface. These diffusing adatoms contribute to both radial and axial growth, hence radial growth competes with axial growth. Because radial growth is kinetically limited, diffusing adatoms are less likely to be incorporated into nanowire sidewalls at lower growth temperatures. Furthermore, adatom diffusion length decreases with decreasing growth temperature. This reduces the flux of adatoms diffusing from the substrate, limiting radial growth and tapering.

The crystal quality of these GaAs nanowires is characterized by TEM as shown in Fig. 2. All nanowires were of ZB structure. Our extensive TEM investigation confirmed that twins and stacking faults exist in the nanowires grown at a high T_g of 450 °C by the single-temperature procedure (Fig. 2a) which is consistent with many previous reports.[12] Twin defects are closely associated with the sidewall faceting behavior of zinc blende GaAs nanowires. Fig. 2a illustrates the nonperiodic sawtooth faceted sidewalls and segmented appearance, typical of nanowires grown at high T_g (by the single-temperature procedure). The faceting occurs in association with the high density of twins in these

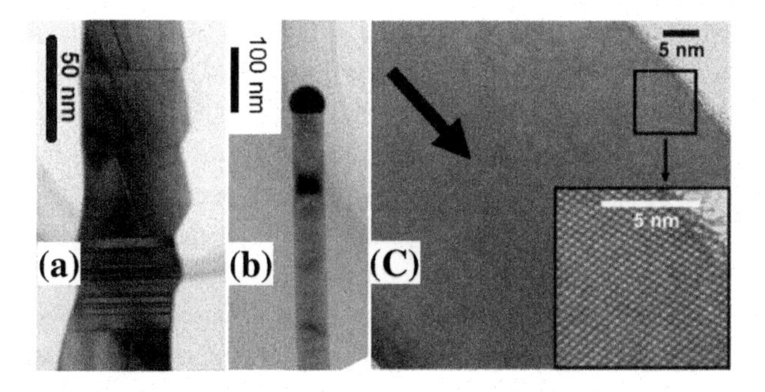

Fig. 2. (a-b) Bright-field TEM and (c) HRTEM images showing morphologies and crystal quality of GaAs nanowires. (a) Nanowires grown at a high T_g of 450 °C by the single temperature procedure, showing twin defects and staking faults. (b-c) Nanowires grown at a low T_g of 390 °C by the two-temperature procedure (T_n of 450 °C), showing no lattice defects.

nanowires: at least one twin defect is associated with the concave facet of each saw-tooth and with each boundary between adjacent segments. Radial growth augments the appearance of sidewall facets. In a remarkable contrast, nanowires grown at lower T_g (by the two-temperature procedure), as illustrated in Fig. 2b-c, exhibit substantially smoother sidewalls. This is related to the elimination of twins and also to the reduction of radial growth.

PL spectra from ensembles of GaAs-AlGaAs core-shell nanowires are plotted in Fig. 3. Nanowire GaAs cores were grown at the higher T_g of 450 °C by the single-temperature procedure, and the lower T_g of 390 °C by the two-temperature procedure. Both samples exhibit the same single broad peak at approximately 1.515 eV. This is consistent with previous measurements on single GaAs nanowires and corresponds closely to excitonic emission in bulk GaAs.[13]

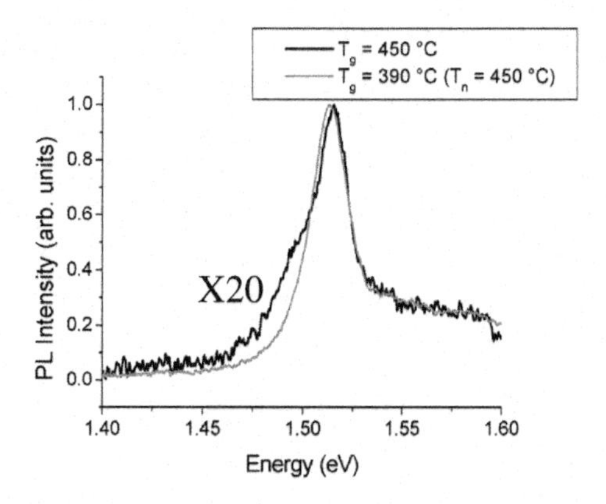

Fig. 3. 10 K PL spectra of GaAs-AlGaAs core-shell nanowires, for two different T_g. GaAs cores were grown at a T_g of 450 °C by the single-temperature procedure (black line, intensity is multiplied by 20 times). GaAs cores were grown at a T_g of 390 °C by the two-temperature procedure with a T_n of 450 °C (gray line).

The PL intensity of 2-temperature grown nanowires is about 20 times higher than the standard 1-temperature grown one, indicating that nanowires grown by the two-temperature process, have superior optical properties to their high-temperature grown counterparts. Improved PL emission is unexpected if we consider the smaller GaAs volume of the thin untapered low-temperature grown nanowires. Three factors could contribute to the observed PL enhancement. Firstly, the low growth temperature eliminates twinning defects, which are believed to adversely affect the optical and electronic properties of the high-temperature grown nanowires.[13] Secondly, the low growth temperature reduces radial overgrowth, which is suspected to be of poor optical quality and may quench photoluminescence. Finally, the irregular facetted sidewalls of nanowires grown at high temperatures (Fig. 2a) may cause roughness at the interface between the GaAs core and AlGaAs shell, resulting in a non-radiative recombination pathway. In contrast, the two-temperature grown nanowires exhibit very smooth sidewalls, leading to a smooth GaAs-AlGaAs interface which reduces non-radiative recombination.

By protecting the AlGaAs shell from oxidation through an additional ~5nm GaAs shell (schematic is shown on the right), we have shown very high optical quality from individual single nanowires with a nearly intrinsic lifetimes (details can be seen in Ref. 14). Similar defect-free GaAs nanowires with high PL emission efficiency have also been achieved by carefully controlling the V/III ratio or growth rate,[15-16] which makes high quality GaAs nanowires possible in various growth conditions.

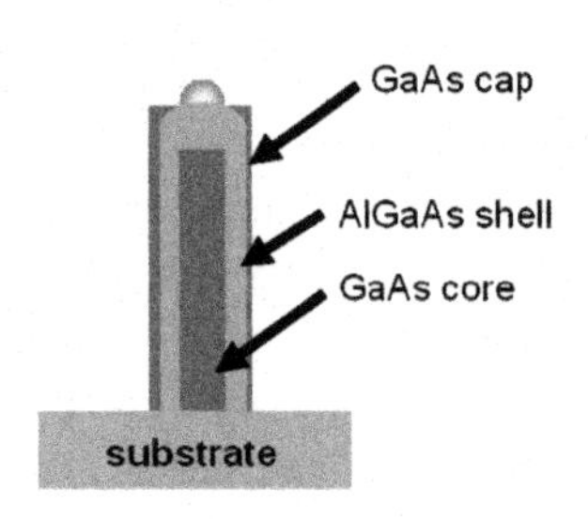

3.2. *GaAs nanowires on Si substrates*

Fig. 4a shows a typical FE-SEM image of GaAs nanowires grown directly on Si (111) substrates, where very low density (~ 0.04 μm^{-2}) of nanowires were observed and exhibit a random growth direction (< 10% of the nanowires are vertical). However, Fig. 4b clearly shows that GaAs buffer layer can dramatically improve the morphology of GaAs nanowires. Nearly no kinked nanowires can be observed. The density and the yield of the vertical nanowires have increased significantly. Although the length of GaAs nanowires grown on Si with GaAs buffer layer is slightly longer than that grown on GaAs (111)B substrates (Fig. 4c), they are almost the same in terms of nanowire morphology and density. The difference in length may be due to the different surface conditions between the GaAs buffer layers (grown on Si substrates) and the GaAs substrates. The cross-sectional SEM image in Fig. 4d clearly showed straight GaAs nanowires grown on Si with a thin buffer layer. TEM image in Fig. 4e illustrates the bottom-part of a nanowire with stacking faults and a saw-teeth shape. Fig. 4f (inset), however, shows perfect ZB structure with the near-hemispherical Au catalyst on tip of the nanowire. Fig. 4g is a high resolution TEM image and shows clearly the ZB structured nanowire with a thin twin

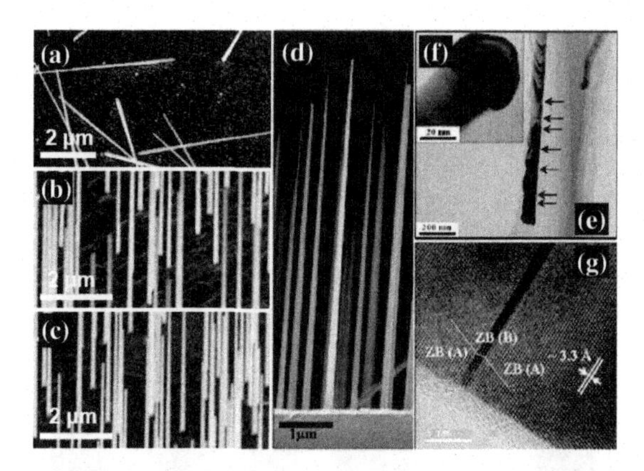

Fig. 4. SEM images (45° tilted) of GaAs nanowires grown on (a) Si (111) substrate directly, (b) Si with GaAs buffer layer, (c) GaAs (111)B substrate directly; (d) Cross-sectional SEM of GaAs nanowires grown on Si (111) substrates with GaAs buffer layer; TEM images of (e) the bottom part of a GaAs nanowire with structural defects, (f) the tip of a GaAs nanowire, and (g) the high-resolution TEM image a GaAs nanowire with stacking faults.

slice (Type A-B-A). Overall, the structural properties of the GaAs nanowires grown on Si wafers with buffer layers are almost same with those standard 1-temperature grown GaAs nanowires on GaAs substrate discussed earlier in this paper. The optical properties of these GaAs nanowires on Si were investigated through micro-PL measurement at 10K. Fig. 5 shows PL spectra from two single nanowires grown on Si and one single nanowires grown on GaAs for comparison. All NWs exhibit nearly identical PL emission at 1.516 eV attributed to free exciton recombination. The PL linewidths are similar for all nanowires, indicating this novel growth of GaAs nanowires on Si is a very promising way to integrate III-V nanowire devices onto a Si platform.

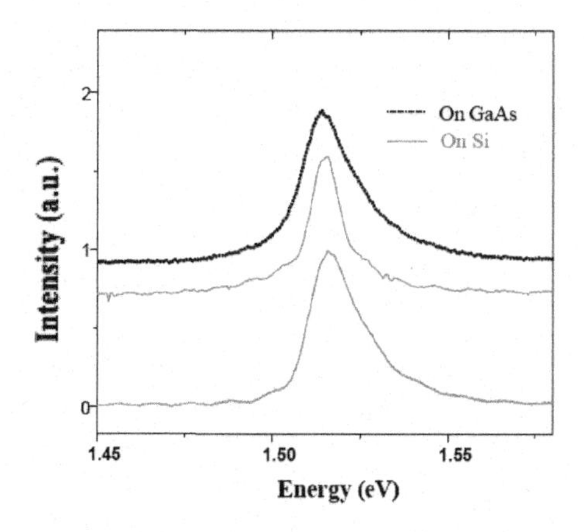

Fig. 5. Micro-PL spectra for a single NW grown on Si substrate (solid line), and GaAs substrate (dotted line). Spectra are shifted for clarity.

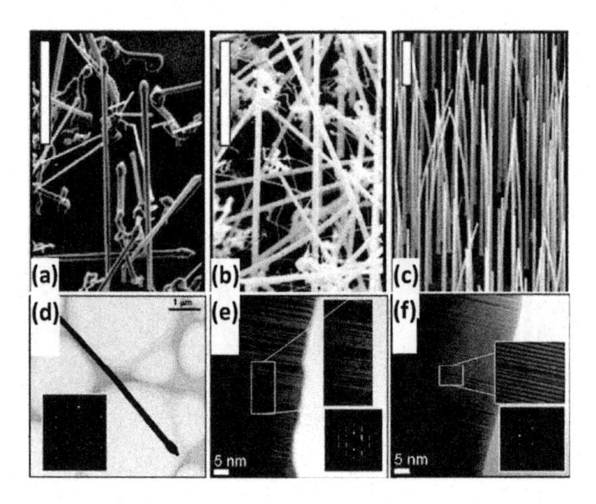

Fig. 6. SEM and TEM images of InP nanowires grown at 400 °C with different V/III ratios (a,d) 110, (b,e) 350 and (c,f) 700. Scale bars in (a-c) are 2 μm.

3.3. *InP nanowires on InP substrates*

SEM (40° tilted) and TEM images of InP nanowires grown at different V/III ratios and catalyzed by 50 nm Au are shown in Fig. 6.[17] Selective area diffraction patterns are slao shown in the inset of TEM images (Figs. 6d-f). Clearly the V/III ratio has a significant effect on the nanowire morphology, which changes from irregular shapes to long and straight vertically aligned nanowires with increasing V/III ratios. The irregular shapes of nanowires at low V/III ratios (up to 350) may suggest insufficient P adatoms due to the incomplete decomposition of PH_3 at such a low growth temperature (400 °C). TEM was carried out to determine the crystal structures of these InP nanowires as shown in Fig. 6d-f. The selected area electron diffraction (SAED) patterns are also shown in insets. At the lowest V/III ratio (110), the SAED pattern clearly shows a perfect ZB structure (Fig. 6d). With increasing V/III ratio, the density of planar defects (such as twins and/or stacking faults) increases significantly. These planar defects create WZ insertions (~40%) in ZB phase, as observed previously. At a V/III ratio of 700, the crystal structure changes to WZ in majority (~ 90% in WZ). We anticipate that the ambient change (i.e. V/III ratio) must have affected the interfacial energy at the growth front and nucleation kinetics, resulting in this alternation in the crystal structures. Additionally, it has been proposed that a high V/III ratio can rapidly quench the growth, causing crystallization in a WZ phase.[18]

By careful control of the V/III ratio and Au nanoparticle size, we are able to obtain InP nanowires which are predominantly a WZ structure with over 150 sections of ZB ranging from 2 to 10 MLs.[19] Fig. 7a is a magnified view of a section of the nanowire where one observes atomic planes of both the WZ and ZB crystal structures. The ZB structured nanowire sections are identified by white dashes labeled with the number of atomic planes included in each section. Fig. 7b shows the band energy diagram and alignments corresponding to this structure. In Fig. 7c & d, we show a series of time

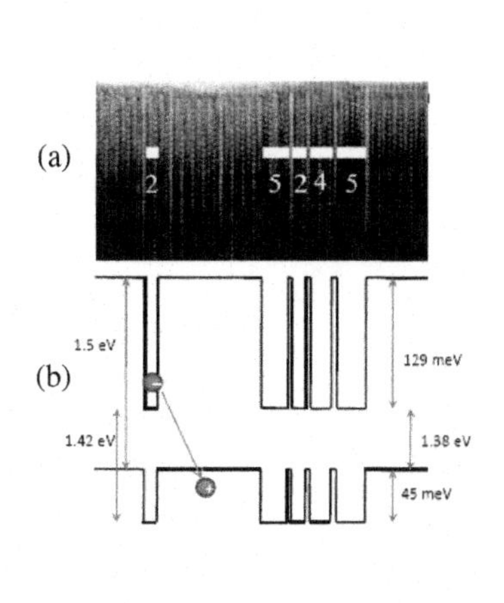

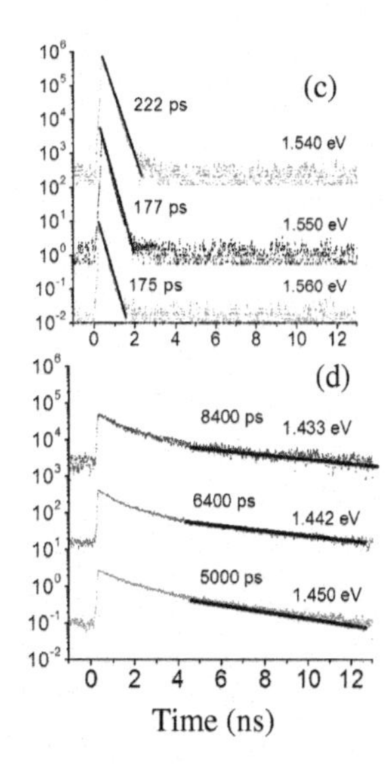

Fig. 7. (a) HRTEM of a nanowire showing the ZB sections identified with the white dashes and the number of atomic planes in a WZ InP nanowire. (b) the band energy diagram and alignments corresponding to the structure indicated in panel a. Time-decays of photoluminescence for energies (c) near the continuum and (d) near the ZB band edge.

decays accumulated from a single nanowire at different energies with the fast MCPT detector and the 250 mm spectrometer. We show time decays of PL in two regions: (i) near the continuum 45 meV (the band offset) above the WZ band WZ band edge or 1.545 eV (Fig. 7c) and (ii) near the ZB band edge at 1.42 eV (Fig. 7d). In region (i), the time decays are single exponential with recombination lifetimes of 175 ps at an energy of 1.56 eV, increasing to 220 ps at 1.54 eV. As the energy moves to region (ii), no significant filling is observed, but the decays are nonexponential at the earliest times: short decays initially, followed by longer decays at later times. At late times, the decay becomes single exponential with a lifetime that increases rapidly with energy up to 8400 ns at 1.433 eV. The fact that lifetimes at the highest energies are an order of magnitude less than this value indicates that the recombination dynamics is most probably limited by capture to lower energy states, while the much longer lifetimes seen at the lowest energy reflect the type-II indirect band alignments between the ZB and WZ sections of InP nanowire. Details of this study can be referred to Ref. 19.

4. Conclusion

In summary, we have reviewed our several research projects on III-V compound semiconductor nanowires. High quality GaAs and InP nanowires and their core/shell or axial hetero crystal structures have been synthesized in a well controlled way. These results show that it is possible to achieve nanowire optical qualities which approach that of the best 2D heterostructures, so that fabrication of highly efficient one dimensional semiconductor nanowire devices and integrate them on Si platform may soon become a reality.

Acknowledgments

This research is supported by the Australian Research Council and Australian National Fabrication Facility established under Australian Government NCRIS Program.

References

1. R.S. Wagner and W.C. Ellis, "Vapor-liquid-solid mechanism of single crystal growth," *Appl. Phys. Lett.* 489-90 (1964).
2. X.F. Duan, Y. Huang, R. Agarwal, and C.M. Lieber, "Single-nanowire electrically driven lasers," *Nature* **421**(6920), 241-245 (2003).
3. M.H. Huang, S. Mao, H. Feick, H.Q. Yan, Y.Y. Wu, H. Kind, E. Weber, R. Russo, and P.D. Yang, "Room-temperature ultraviolet nanowire nanolasers," *Science* **292**(5523), 1897-1899 (2001).
4. J.F. Wang, M.S. Gudiksen, X.F. Duan, Y. Cui, and C.M. Lieber, "Highly polarized photoluminescence and photodetection from single indium phosphide nanowires," *Science* **293**(5534), 1455-1457 (2001).
5. J. Hahm and C.M. Lieber, "Direct ultrasensitive electrical detection of DNA and DNA sequence variations using nanowire nanosensors," *Nano Lett.* **4**(1), 51-54 (2004).
6. R.E. Algra, M.A. Verheijen, M.T. Borgstrom, L.F. Feiner, G. Immink, W.J.P. van Enckevort, E. Vlieg, and E. Bakkers, "Twinning superlattices in indium phosphide nanowires," *Nature* **456**(7220), 369-372 (2008).
7. P. Caroff, K.A. Dick, J. Johansson, M.E. Messing, K. Deppert, and L. Samuelson, "Controlled polytypic and twin-plane superlattices in III–V nanowires," *Nature Nanotechnol.* **4**50-55 (2009).
8. J. Zou, M. Paladugu, H. Wang, G.J. Auchterlonie, Y.N. Guo, Y. Kim, Q. Gao, H.J. Joyce, H.H. Tan, and C. Jagadish, "Growth mechanism of truncated triangular III-V nanowires," *Sml* **3**(3), 389-393 (2007).
9. J.H. Kang, Q. Gao, H.J. Joyce, H.H. Tan, C. Jagadish, Y. Kim, D.Y. Choi, Y. Guo, H. Xu, J. Zou, M.A. Fickenscher, L.M. Smith, H.E. Jackson, and J.M. Yarrison-Rice, "Novel growth and properties of GaAs nanowires on Si substrates - art. no. 035604," *Nanotechnology* **21**(3), 35604.
10. H.J. Joyce, Q. Gao, H.H. Tan, C. Jagadish, Y. Kim, X. Zhang, Y.N. Guo, and J. Zou, "Twin-free uniform epitaxial GaAs nanowires grown by a two-temperature process," *Nano Letters* **7**(4), 921-926 (2007).
11. J. Johansson, C.P.T. Svensson, T. Martensson, L. Samuelson, and W. Seifert, "Mass transport model for semiconductor nanowire growth," *J. Phys. Chem. B* **109**(28), 13567-13571 (2005).
12. A. Mikkelsen, N. Skold, L. Ouattara, M. Borgstrom, J.N. Andersen, L. Samuelson, W. Seifert, and E. Lundgren, "Direct imaging of the atomic structure inside a nanowire by scanning tunnelling microscopy," *Nat. Mater.* **3**(8), 519-523 (2004).
13. L.V. Titova, T.B. Hoang, H.E. Jackson, L.M. Smith, J.M. Yarrison-Rice, Y. Kim, H.J. Joyce, H.H. Tan, and C. Jagadish, "Temperature dependence of photoluminescence from single core-shell GaAs-AlGaAs nanowires," *Appl. Phys. Lett.* **89**(17), 73126 (2006).
14. S. Perera, M.A. Fickenscher, H.E. Jackson, L.M. Smith, J.M. Yarrison-Rice, H.J. Joyce, Q. Gao, H.H. Tan, C. Jagadish, X. Zhang, and J. Zou, "Nearly intrinsic exciton lifetimes in single twin-free GaAs/AlGaAs core-shell nanowire heterostructures," *Appl. Phys. Lett.* **93**(5), 53110 (2008).

15. H.J. Joyce, Q. Gao, H.H. Tan, C. Jagadish, Y. Kim, M.A. Fickenscher, S. Perera, T.B. Hoang, L.M. Smith, H.E. Jackson, J.M. Yarrison-Rice, X. Zhang, and J. Zou, "High purity GaAs nanowires free of planar defects: growth and characterization," *Adv. Funct. Mater.* **18**3794-3800 (2008).
16. H.J. Joyce, Q. Gao, H.H. Tan, C. Jagadish, Y. Kim, M.A. Fickenscher, S. Perera, T.B. Hoang, L.M. Smith, H.E. Jackson, J.M. Yarrison-Rice, X. Zhang, and J. Zou, "Unexpected benefits of rapid growth rate for III-V nanowires," *Nano Letters* **9**(2), 695-701 (2009).
17. S. Paiman, Q. Gao, H.H. Tan, C. Jagadish, K. Pemasiri, M. Montazeri, H.E. Jackson, L.M. Smith, J.M. Yarrison-Rice, X. Zhang, and J. Zou, "The effect of V/III ratio and catalyst particle size on the crystal structure and optical properties of InP nanowires - art. no. 225606," *Nanotechnology* **20**(22), 25606 (2009).
18. K.A. Dick, K. Deppert, L.S. Karlsson, M.W. Larsson, W. Seifert, L.R. Wallenberg, and L. Samuelson, "Directed growth of branched nanowire structures," *MRS Bulletin* **32**127-133 (2007).
19. K. Pemasiri, M. Montazeri, R. Gass, L.M. Smith, H.E. Jackson, J. Yarrison-Rice, S. Paiman, Q. Gao, H.H. Tan, C. Jagadish, X. Zhang, and J. Zou, "Carrier dynamics and quantum confinement in type II ZB-WZ InP nanowire homostructures," *Nano Letters* **9**(2), 648-654 (2009).

ELECTRON HEATING IN QUANTUM-DOT STRUCTURES WITH COLLECTIVE POTENTIAL BARRIERS

L.H. CHIEN

EE Department, University at Buffalo, The State University of New York, 321 Bonner Hall, Buffalo, NY 14260, USA
lchien2@buffalo.edu

A. SERGEEV

SUNY Research Foundation, University at Buffalo, The State University of New York, 320 Bonner Hall, Buffalo, NY 14260, USA
asergeev@eng.buffalo.edu

N. VAGIDOV

EE Department, University at Buffalo, The State University of New York, 312 Bonner Hall, Buffalo, NY 14260, USA
nizami@buffalo.edu

V. MITIN

EE Department, University at Buffalo, The State University of New York, 312 Bonner Hall, Buffalo, NY 14260, USA
vmitin@buffalo.edu

S. BIRNER

Physics Department, Technische Iniversitat Munchen, Am Coulombwall 3, D-85748 Garching, Germany
stefan.birner@wsi.tum.de

Here we report our research on quantum-dot structures with collective barriers surrounding groups of quantum dots (planes, clusters etc) and preventing photoelectron capture. Employing Monte-Carlo simulations, we investigate photoelectron kinetics and calculate the photoelectron lifetime as a function of geometrical parameters of the structures, dot occupation, and electric field. Results of our simulations demonstrate that the capture processes are substantially suppressed by the potential barriers and enhanced in strong electric fields. Detailed analysis shows that the effects of the electric field can be explained by electron heating, i.e. field effects become significant, when the shift of the electron temperature due to electron heating reaches the barrier height. Optimized photoelectron kinetics in quantum-dot structures with collective barriers allows for significant improvements in the photoconductive gain, detectivity, and responsivity of photodetectors based on these structures.

Keywords: quantum-dot photodetector; potential barriers; photoelectron lifetime; capture; gain.

1. Introduction

Adequate understanding of photoelectron kinetics in nanostructures is critically important for attainment of high performance in room-temperature semiconductor optoelectronic devices, such as mid- and far-infrared detectors, solar cells, etc.[1-6] During many years quantum-dot nanostructures were considered as a promising candidate for improving the room-temperature operation due to expected slow relaxation between discrete QD levels. These expectations were based on the "phonon bottleneck" concept, which assumes that the phonon-assisted bound-to-bound transitions are prohibited, unless the energy between two discrete levels matches the phonon energy.[7] According to this concept, the intrinsic electron relaxation in QDs was anticipated to be significantly slower than that in 2D and 3D structures. However, the phonon bottleneck model completely ignores modification of electron states due to interaction effects, e.g. due to a finite width of electron energy levels.

Recent investigations[1] unambiguously demonstrated that the actual intra-dot kinetics is completely opposite to what can be expected for weakly interacting electrons and phonons. In reality, strong coupling between electrons and longitudinal optical (LO) phonons leads to formation of the polaron states, which decay due to the interaction of LO phonons with acoustical phonons. Such kinetics results in strong energy and temperature dependences of the electron relaxation. At helium temperature, long relaxation time (~ 1.5 ns) was observed for the level separation of 14 meV (3.4 THz).[1] However, the relaxation time decreases to ~ 2 ps for the 30 meV transition. The relaxation time also drastically decreases, if temperature increases. For example, for 14 meV transition, the relaxation time was reduced from 1.5 ns at 10 K to 560 ps at 30 K, and further to 260 ps at 50 K. At room temperatures the polaron decay time is observed in the range of 2 – 30 ps, depending on the electron energy.[1] After numerous experiments[1,8-10] with various QD structures, no true phonon bottleneck has been found. Thus, the intra-dot electron kinetics at room temperatures turns out to be very fast and practically unmanageable.

In our previous publications[11-15] we proposed to manage inter-dot kinetics and suppress photocarrier capture processes by means of the barriers, which are formed by electrons bounded in quantum dots and ionized impurities in the depletion region. These potential barriers separate the conducting electron states from the localized intra-dot states and prevent photoelectron capture into the dots. Modern technologies provide many possibilities to fabricate QD infrared photodetectors (QDIPs) with local and collective potential barriers. The local potential barrier around a single dot can be formed by homogeneous doping of the interdot space. To suppress electron capture process, the barrier height should be at least two - three times larger than kT. Therefore, at room temperatures, the local barriers should be ~ 0.1 eV and quantum dots should comprise at least ten electrons. This requires large size dots and very high level of doping.

At the same time, significant potential barriers that effectively separate conducting and localized states can be created by groups of quantum dots (dot clusters, rows etc).[16] Such collective barriers divide the groups of quantum dots from high-mobility conducting channels. In this work we report results of our modeling of photoelectron kinetics and transport in QD structure with collective potential barriers.

2. Nanostructures with Collective Barriers

The QD structures with collective potential barriers provide new possibilities to suppress capture processes and control photocarrier kinetics. Fig. 1.a shows the QD structure with the lateral transport along high-mobility heterointerfaces, Fig. 1.b demonstrates the QD structure with vertically correlated dot clusters (VCDC) and the vertical transport through inter-cluster areas. In a QD structure with the lateral transport (Fig. 1.a) the potential barriers are formed by charged QD planes and charged planes of dopants. In VCDC structures (Fig. 1.b) the barriers are created by charged dot clusters.

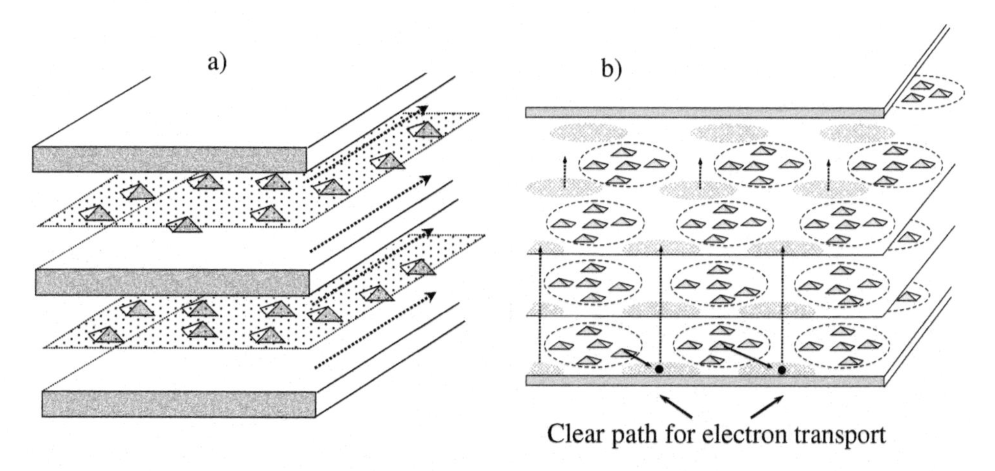

Fig. 1. (a). Schematic view of QD structures with collective potential barriers, which separate QD areas from the areas transmitting the photocurrent: (a) Structure with lateral transport (dashed arrows) along 2D heterointerfaces, (b) Structure with vertically correlated QD clusters and vertical transport in the inter-cluster areas. QDs are separated from conducting channels by potential barriers, which suppress photoelectron capture into QDs.

For example, the photodetector with the lateral transport (Fig. 1.a) may be realized on the basis of modulation-doped AlGaAs/GaAs structures with self-assembled InAs dots. QD planes are imbedded in the middle of GaAs quantum well with a thickness $2d$. AlGaAs layers are doped to supply free carriers. This lateral QDIP conducts photocurrent through high-mobility two-dimensional channels along the heterointerfaces of AlGaAs/GaAs. The potential barriers are formed by the ionized dopants in AlGaAs and the charges in QDs and have the form of $V_m = e^2 N_d^{(2)} nd/(2\varepsilon\varepsilon_0)$, where $N_d^{(2)}$ is the dot concentration in QD planes and n is the average occupation in the dots.

The detector based on the structures with the vertically correlated dot clusters is shown in Fig. 1.b. Positions of the dot clusters are correlated in the vertical direction, which is the direction of photocurrent. The collective barriers around dot clusters are created by charged dots and ionized dopants. This structure has the advantage of the vertical QDIP, which utilizes the mature and well-established technology for Ohmic contacts. In both structures, the electrons can be directly excited to the conducting states or excited to the quasi-localized high-energy states and then emitted to the conducting states via thermoexcitation, tunneling, or photon-assisted tunneling. In VCDC structures the photoelectrons move in the areas between dot clusters through the high-mobility channels. If the radius of the cluster, b, exceeds the distance between dot layers, c, the potential barrier around dot clusters has a logarithmic form of $V_m = e^2Nn/(2\pi\varepsilon\varepsilon_0 c)\ \ln(w/b)$, where N is the number of dots in the cluster, n is the average number of carriers per dot, and $2w$ is the distance between the centers of two neighboring clusters, so the dot cluster concentration is $N_c = N_d^{(2)}/N$, where $N_d^{(2)}$ is the dot concentration in the QD planes.

The self-consistent 3D-simulations of VCDC structures using nextnano3 software developed at the Walter Schottky Institut,[17] gives the spatial distribution of potential profile with barriers surronding individual QDs as well as around the cluster of QDs (see Fig. 2. Cluster consisting of 9 InAs QDs in GaAs matrix is shown encircled).

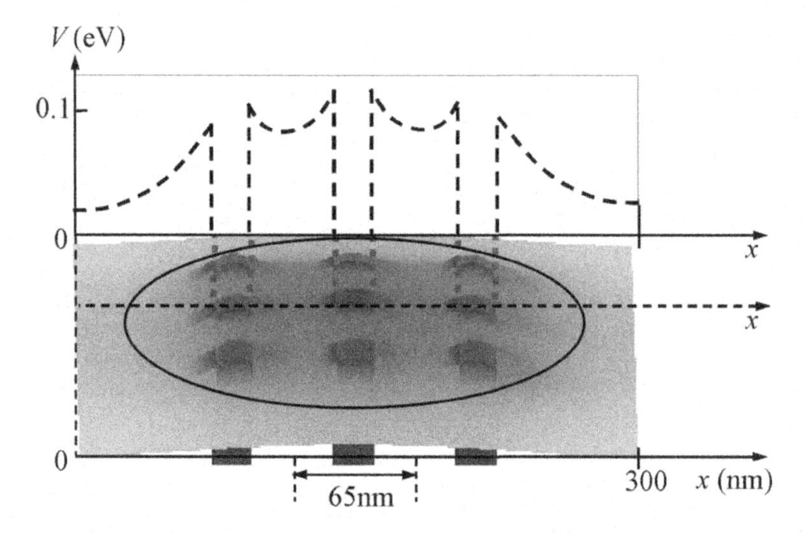

Fig. 2. The potential profile of the layer of VCDC structure with 9 QDs per cluster. The dimensions of InAs QDs in GaAs matrix are 20×20×10 nm^3, QD areal density is 10^{10} cm^{-2}, and doping is 4×10^{10} cm^{-2} (4 electrons per QD).

The simulated QD structure consisted of 5 interchanging layers with QD clusters and layers of GaAs which were intentionally doped. Fig. 2 shows both three-dimensional potential profile as well as its cross-section in the middle of QD cluster shown by the dashed line. Dark area represent higher potential barriers whereas lighter area - low potential values. It is seen from Fig. 2 that in the area, which is outside of the QD cluster, the electron potential energy is about 10 meV, whereas around QDs it is about 110 meV slightly varying between QDs. Thus, in the area that is far from the QD cluster electrons can freely move laterally as well as in the perpendicular to the QD cluster layer direction. The characteristic picture of the QD cluster potential profile was obtained as it is seen from Fig. 2 for the QD areal density equal to 10^{10} cm^{-2}. The doping of the GaAs layer adjacent to the layer with QD clusters is equal to 4×10^{10} cm^{-2}, thus resulting in ~4 electrons per each QD.

The photoelectron lifetime, i.e. the photocarrier capture time in QD structures, is a critical parameter of QDIP. The limiting values of the noise equivalent power, NEP, and detectivity, D^*, are determined by the generation - recombination noise, which is controlled by the capture processes [3],

$$NEP = \frac{h\nu}{\eta}\sqrt{\frac{2n_{th}V}{\tau_{cap}}}, \quad D^* \equiv \frac{\sqrt{S}}{NEP_{GR}} = \frac{\eta}{h\nu}\sqrt{\frac{\tau_{cap}}{2n_{th}d}}, \qquad (1)$$

where η is the total quantum efficiency, n_{th} is the density of the thermally activated electrons in conducting states, V, S, and d are the sensor volume, area, and width, correspondingly. In weak electric fields, the capture rate, τ_{cap}^{-1}, may be calculated analytically. Calculations show that it depends exponentially on the value of the barrier height, V_m, [3]

$$\frac{1}{\tau_{capt}} = \frac{1}{\tau_0}\exp\left(-\frac{eV_m}{kT}\right), \qquad (2)$$

where τ_0 is the capture time in the flat potential structure with the same geometrical parameters. The parameter τ_0 depends on the dot concentration, N_d, the characteristic dot size, a, with respect to the electron mean free path, ℓ, and positions of quantum dots.

According to Eq. 2, the capture time in the lateral structures exponentially increases with increasing of the dot concentration, dot's occupation, and the GaAs width $2d$. In the VCDC structures, the capture time exponentially increases with increasing of the number of dots in a cluster and dot's occupation. Because of the logarithmic dependence of the potential barriers on the geometrical parameters, the capture time depends weakly on the cluster radius and average distance between dot clusters.

Analytical consideration of photoelectron kinetics is limited by small electric fields. To minimize the photoelectron transit time and to increase the photoconductive gain, QDIPs operate at significant biased voltages, which substantially change the photoelectron distribution functions. In the next section we consider kinetics in the lateral QD structures (Fig. 1.a) and VCDC structures (Fig. 1.b) on the basis of the Monte-Carlo modeling.

3. Photoelectron Kinetics in Strong Electric Fields

Our Monte-Carlo program includes all basic scattering mechanisms, such as electron scattering on acoustic, polar optical, and intervalley phonons. The program considers electrons which may populate Γ–, L–, and X– valleys and takes into account redistribution of carriers between valleys created by charged dots. We consider the carrier capture process as a specific inelastic scattering process, which is confined in space by the dot volume and in which a carrier is transferred from a conducting state above the potential barrier to a bound state below the barrier. We assume that from a bound state a carrier will relax to the deep dot states faster than it could return back to the conducting state.

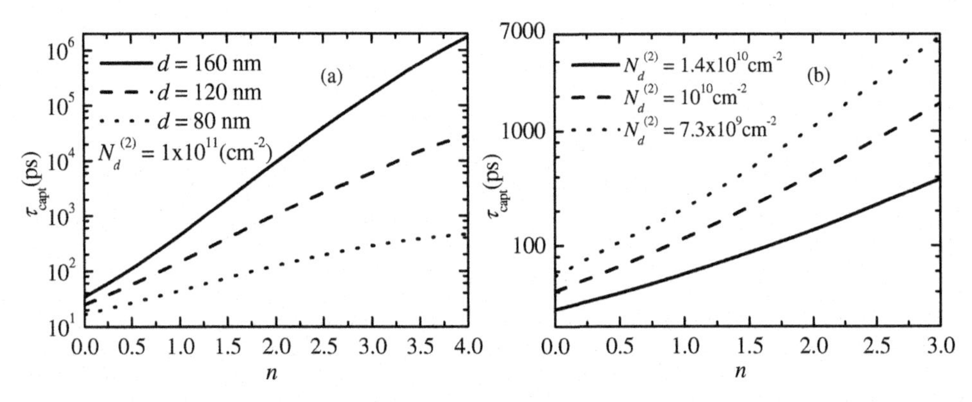

Fig. 3. Capture time as a function of potential barrier height, V_m, in QD structures (a) with lateral transport and (b) with VCDC and vertical transport.

Figure 3 presents the dependence of the capture time on the dot occupation, n, for both the lateral and VCDC structures. Parameters of the lateral structures are shown in the figure. For the VCDC structures we take the following parameters: $N = 9$, $b = 75$ nm, $c = 40$ nm, and the distance between the neighboring dots is 55 nm. As seen, the capture time has an exponential dependence on occupation, as it is expected from Eq. 2 taking into account that the barrier height depends linearly on the occupation.

Figure 4 shows the capture time as a function of the dot concentration. The dot concentration is an important factor to optimize the absorption of radiation by QD structures. Modern technologies can provide structures with dot concentration in the range of 10^9 - 10^{11} cm^{-2}. In the lateral structures, the potential barrier height increases linearly with increasing dot concentration. As seen this effect prevails at the dot occupation n = 2 and 3. At n =1, the effect has nonmonotonic character. At small dot concentations the barriers are small and by increasing the dot concentration we increase the number of traps for photoelectrons. Therefore, the capture time decreases with increasing the dot concentration. Charged dots form substantial potential barriers and the photoelectron time increases with increasing the dot concentration.

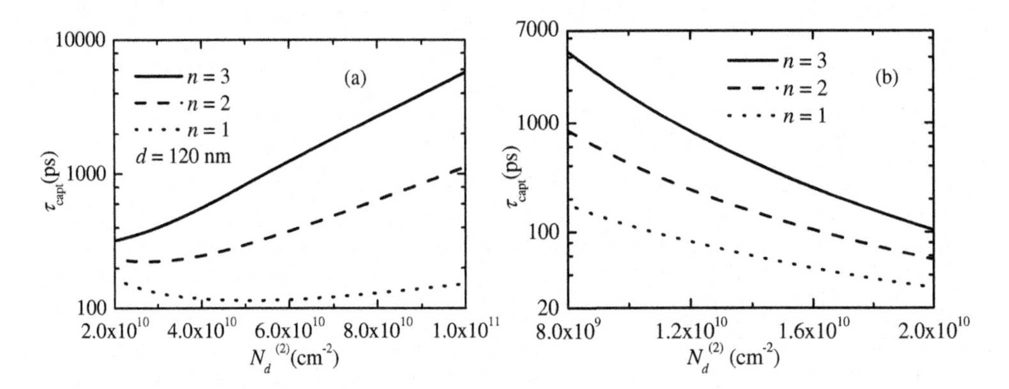

Fig. 4. Capture time as a function of dot concentration, $N_d^{(2)}$, in QD structures (a) with lateral transport and (b) with VCDC and vertical transport.

In VCDC structure, the dependence of capture time on dot concentration is opposite to that in the lateral structures (Fig. 4.b). In the VCDC structures, an increase of the dot concentration at fixed number of the dots per cluster leads to the decrease of the potential barriers. Besides this, it also results in an increase of traps for photoelectrons. For these reasons, capture time decreases with the increasing of the dot concentration.

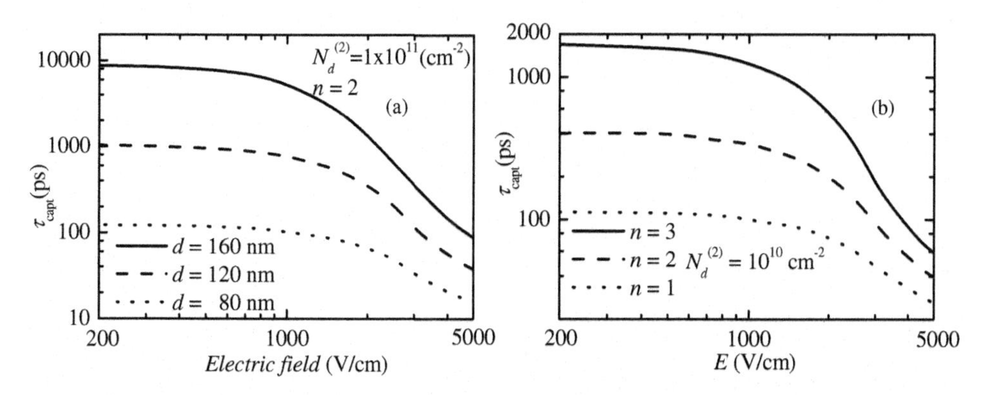

Fig. 5. Capture time as a function of electric field in QD structures (a) with lateral transport and (b) with VCDC and vertical transport.

Figure 5 shows the capture time as a function of electric field for the lateral and VCDC structures. At small electric field, the capture time changes just slightly and remains almost constant up to a characteristic value of electric field, ~ 1000 V/cm. Further increase in the electric field leads to substantial enhancement of the capture processes.

To understand the effects of the electric field on the capture time, we calculated the dependence of the average electron energy $\bar{\varepsilon}$ on the electric field (Fig. 6). Let us note that the average electron energy may be also evaluated from the energy balance equation, if one assumes that the nonequilibrium electron distribution function is described by

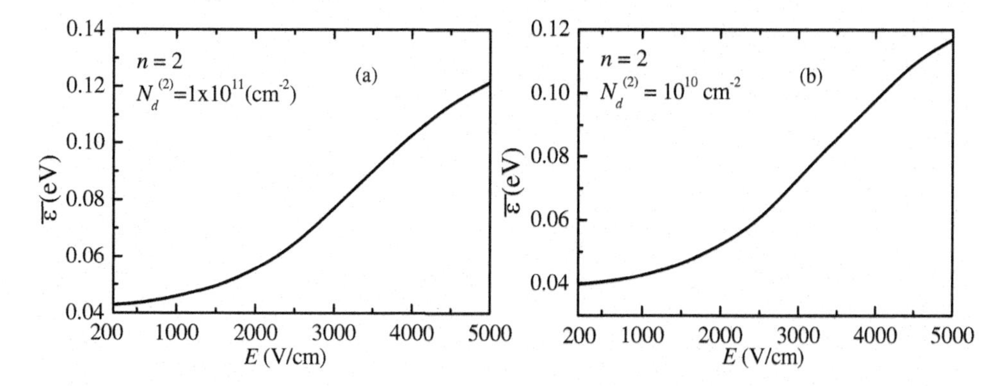

Fig. 6. Average electron energy as a function of the electric field in QD structures (a) with lateral transport and (b) with VCDC and vertical transport.

electron temperature. Both numerical and analytical calculations provide consistent results. Note, that in this range of field, the potential barriers don't change this dependence.

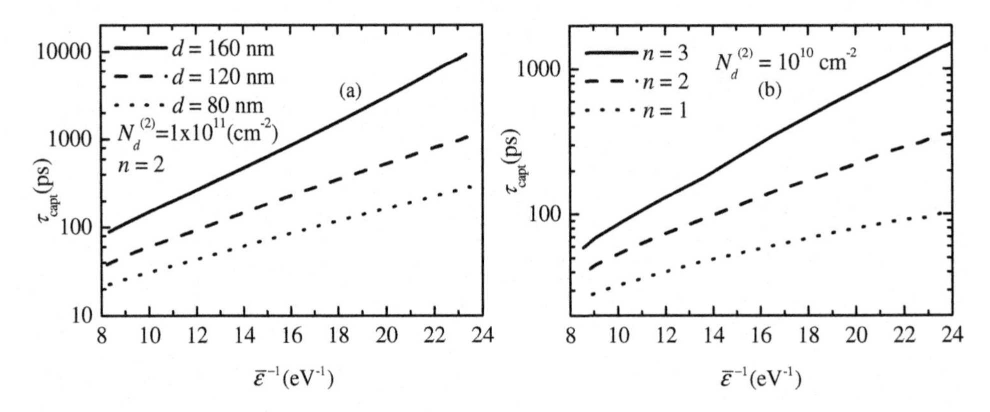

Fig. 7. Capture time as a function of the inverse electron energy attained in the electric field in QD structures (a) with lateral transport and (b) with VCDC and vertical transport.

Figure 7 presents the dependence of the capture time on the inverse value of $\bar{\varepsilon}$. As it is seen, log τ_{capt} is proportional to $1/\bar{\varepsilon}$, i.e. τ_{capt} is proportional to $\exp(1/\bar{\varepsilon})$. Thus, the carrier capture in the electric field can be described by Eq. 2, where the thermal energy kT is replaced by a factor of $\sim 2\bar{\varepsilon}/3$. Thus, we may conclude that the effect of the electric field on carrier capture is well described by the model of electron heating.

Finally, we calculated the photoconductive gain, g, as a function of the electric field for devices with the length of 1 μm. Photoconductive gain is defined as the ratio of the carrier lifetime, τ_{cap}, to carrier transit time, τ_{tr}. The photoelectron transit time is the time that the electron spends in the device moving from the emitter to the collector and, therefore, τ_{tr} is inversely proportional to the drift velocity. The average drift velocities have been calculated using the same Monte Carlo program. As seen from Fig. 8, the gain

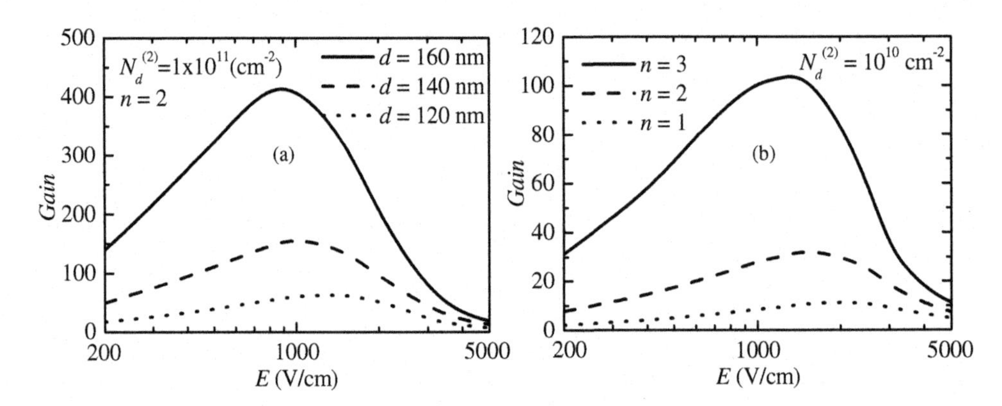

Fig. 8. Photoconductive gain as a function of electric field in QD structures (a) with lateral transport and (b) with VCDC and vertical transport. Length of both devices is 1 μm.

approaches a maximum value at electric field of the order of 10^3 V/cm, which is also the characteristic field for the dependences shown in Fig. 5. This nonmonotonic dependence on the electric field may be explained in the following way. At small electric fields, the gain increases with increasing of the electric field, since the transit time reduces and capture time remains almost constant. When the electric field increases up to a characteristic value, transit time almost saturates and capture time reduces substantially and, therefore, the gain decreases substantially.

In summary, photodetectors based on quantum dot structures with specially engineered potential barriers have a strong potential to overcome limitations of quantum-well-based devices for IR imaging. Such structures have a number of characteristics making them especially suitable for optoelectronic applications: (i) Manageable photoelectron kinetics; (ii) High photoconductive gain and responsivity; (iii) Low generation-recombination (GR) noise due to low concentration of thermo-excited carriers; (iv) High scalability of nanoblocks and numerous possibilities for nano-engineering; (v) Available fabrication technologies.

Here we presented the modeling of the photoelectron kinetics in the lateral structures and structures with vertically correlated dot clusters. Our results show that collective potential barriers around the groups of QDs effectively prevent photoelectron capture into the dots, while electric field overheats photoelectrons and enhances capture processes. Interplay of these two factors determines the optimal operating regime for QDIPs with potential barriers.

Acknowledgements

The research was supported by AFOSR (Mitin and Sergeev) and NSF under Grant No DMR 0907126 (Chien).

References

1. E. A. Zibik, T. Grange, B. A. Carpenter, N. E. Porter, R. Ferreira, G. Bastard, D. Stehr, S. Winnerl, M. Helm, H. Y. Liu, M. S. Skolnick, and L. R. Wilson, *Nature Materials* **8**, 403 (2009).
2. M. Razeghi, H. Lim, S. Tsao, J. Szafraniec, W. Zhang, K. Mi, and B. Movaghar, *Nanotechnology* **16**, 219 (2005).
3. J. C. Campbell and A. Madhukar, *IEEE Quantum Electronics* **95**, 1815 (2007).
4. A. V. Barve, S. J. Lee, S. K. Noh, and S. Krishna, *Laser & Photonics Reviews*, June 17 (2009).
5. M. D. Kelzenberg, D. B. Turner-Evans, B. M. Kayes, M. A. Filler, M. C. Putnam, N. S. Lewis, H. A. Atwater et al., *NANO Letters* **8**, 710 (2008).
6. H. Spanggaard and F. C. Krebs, *Solar Energy Materials and Solar Cells* **83**, 125 (2004).
7. U. Bockelmann and G. Bastard, *Phys. Rev. B* **42**, 8947 (1990).
8. Y. Toda, O. Moriwaki, M. Nishioka, and Y. Arakawa, *Phys. Rev. Lett.* **82**, 4114 (1999).
9. R. Ferreira and G. Bastard, *Appl. Phys. Lett.* **74**, 2818 (1999).
10. H. Lim, B. Movaghar, S. Tsao, M. Taguchi, W. Zhang, A. A. Quivy, and M. Razeghi, *Phys. Rev. B* **74**, 205321 (2006).
11. A. Sergeev, V. Mitin, and M. Stroscio, *Physica B* **316-317**, 369-372 (2002).
12. V. Mitin, N. Vagidov, and A. Sergeev, *Physica Status Solidi (C)* **3**, 4013-4016 (2006).

13. V. Mitin, A. Sergeev, L. H. Chien, and N. Vagidov, *Proceed. of SPIE: Nanophotonics and Macrophotonics for Space Environment II*, **7095**, 70950K1 - 70950K9 (2008).
14. L.-H. Chien, A. Sergeev, N. Vagidov, and V. Mitin, *Int. J. of High Speed Electronics and Systems* **18**, 1013-1022 (2008).
15. N. Vagidov, A. Sergeev, and V. Mitin, *Physics and Modeling of Tera- and Nano-Devices*, editors M. Ryzhii and V. Ryzhii, World Scientific, 141-148 (2008).
16. L. H. Chien, A. Sergeev, V. Mitin, and S. Oktyabrsky, *Proceed. of SPIE: SPIE Photonics West, Connecting Minds for Global Solutions* (2010), will be published.
17. http://www.nextnano.de/nextnano3/

ELECTRONIC STRUCTURE OF GRAPHENE NANORIBBONS SUBJECTED TO TWIST AND NONUNIFORM STRAIN

A. DOBRINSKY

Engineering Department, Brown University,
182 Hope St. Providence, RI 02906, USA
alex.dobrinsky@gmail.com

A. SADRZADEH

Mechanical Engineering Department, Rice University,
6100 Main St. Houston TX 77006, USA
sadrzadeh0@yahoo.com

B. I. YAKOBSON

Mechanical Engineering Department, Rice University,
6100 Main St. Houston TX 77006, USA
biy@rice.edu

J. XU

Engineering Department, Brown University,
182 Hope St. Providence, RI 02906, USA
Jimmy_Xu@Brown.Edu

Graphene nanoribbons exhibit band gap modulation when subjected to strain. While band gap creation has been theoretically investigated for uniaxial strains, other deformations such as nanoribbon twist have not been considered. Our main objective in this paper is to explore band gap opening in twisted graphene nanoribbons that have metallic properties under tight-binding approximation. While simple considerations based on the Hückel model allow to conclude that zigzag graphene nanoribbons exhibit no band gap when subjected to twist, the Hückel model overall may be inaccurate for band gap prediction in metallic nanoribbons. We utilize Density Functional Theory Tight-Binding Approximation together with a requirement that energy of twisted nanoribbons is minimized to evaluate band gap of metalic armchair nanoribbons. Besides considering twisting deformations, we also explore the possibility of creating band gap when graphene nanoribbons are subject to inhomogeneous deformation such as sinusoidal deformations.

Keywords: graphene nanoribbons; band structure.

1. Introduction

Graphene has attracted a great amount of interest due to its unusual electronic properties[1,2]. Graphene nanoribbons electronic structure can be well tuned by varying the width of graphene ribbons through e-beam lithography[3,4]. Recently it was shown that electronic structure of graphene thin films, such as bilayer graphene[5], as well as graphene nanoribbons (GNR)[6,7] were influenced by electric and magnetic fields. Further the influence of uniform deformations on electronic properties was studied using tight-binding formulation[8] followed by first principles[9] calculations. At the same time, effects of unequal hopping parameters between neighboring graphene atoms were analyzed in the tight-binding framework[10] in order to understand how variability of hopping parameters can lead to the possibility of band gap opening in graphene. It was found that graphene is robust to uniform strain deformations, and acquires band gap only for strains of larger then 23%[11]. While graphene band gap modulation due to stretching was investigated[11], the band gap modulation due to spontaneous twisting of graphene nanoribbons has not been studied. Since it has been shown[12,13] that graphene nanoribbons can undergo spontaneous twist, studying electronic structure of graphene nanoribbons subject to twist is an interesting and relevant topic. Besides spontaneous twist, graphene nanoribbons exhibit other instabilities, such as waviness. Ripple effect has been recognized in graphene and was compared to "pseudo magnetic field"[14]. While ripple instabilities are present in large graphene sheets, they are also a major instability in graphene nanoribbons[12]. In this paper we consider effects of twist and sinusoidal waviness of GNR on their electronic structure by studying both armchair and zigzag nanoribbons. Since our main objective is to explore band gap opening we have considered only ribbons that have metalic properties under tight-binding approximation. This includes zigzag ribbons and armchair ribbons with $N = 3k + 2$, where N is the number of armchair chains[15].

2. Tight-Binding Formulation, Hückel Mode

To get qualitative understanding of twisting and stretching effects on band gap characteristics of GNR we use the Hückel model, and first, apply it to zigzag nanoribbons. A typical unit cell of zigzag nanoribbon is illustrated in Fig. 1. The Hamiltonian matrix is given by:

$$H = \begin{bmatrix} \alpha_1 & \beta_1 & \cdots & 0 & 0 \\ \tilde{\beta}_2^T & \alpha_2 & \beta_2 & 0 & 0 \\ \vdots & \ddots & \ddots & \ddots & \vdots \\ 0 & 0 & \ddots & \ddots & \beta_m \\ 0 & 0 & 0 & \tilde{\beta}_m^T & \alpha_m \end{bmatrix} \tag{1}$$

$$\alpha_i = \begin{pmatrix} 0 & t_{12}^i + t_{12N}^i e^{-ika} & 0 & 0 \\ \tilde{t}_{12}^i + \tilde{t}_{12N}^i e^{ika} & 0 & t_{23}^i & 0 \\ 0 & \tilde{t}_{23}^i & 0 & t_{34}^i + t_{34N}^i e^{ika} \\ 0 & 0 & \tilde{t}_{34}^i + \tilde{t}_{34N}^i e^{-ika} & 0 \end{pmatrix} \tag{2}$$

$$\beta_{ij} = t_{ij}\delta_{i4}\delta_{j1}, i, j \in \{1..4\} \tag{3}$$

Here, a is a lattice constant, and $\tilde{t}$ denotes a complex conjugate of hopping parameter t. Subscript N denotes hopping between atom in a unit cell and its neighbor in a neighboring unit cell. Index i runs over all of the "group units" shown in Figure $1(a)$.

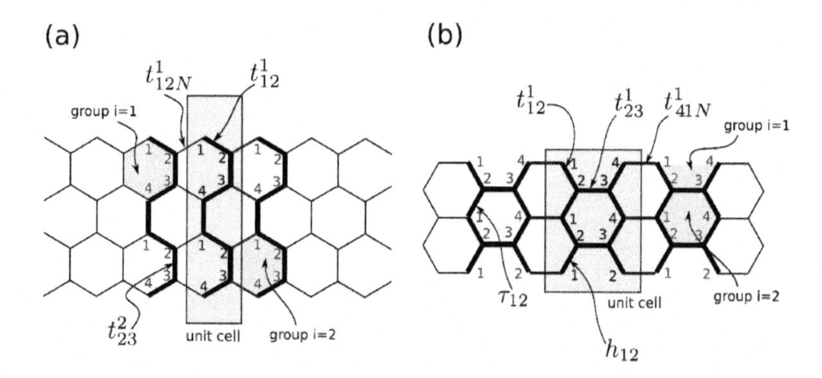

Fig. 1. Schematics of a unit cell in a zigzag and armchair nanoribbons.

Note that under uniform axial deformations the hopping parameters are

$$t^i_{12} = t^i_{12N}, t^i_{34} = t^i_{34N} \tag{4}$$

Therefore at Briluoin zone boundaries, $k = \pm\pi / a$, Hamiltonian of the system contains rows and columns of zeros that results in several eigenvalues being zero. How that translates in the absence of band gap is not immediately apparent, but under assumption of symmetry of valence and conduction bands, this results in lack of a band gap. We have confirmed the lack of the band gap by varying hopping parameters and calculating band gap structure based on Hückel model.

Note, that this result is independent of values of t^i_{23}. The absence of the band gap for zigzag nanoribbons under uniaxial stretching has been reported in the literature[8,9] and does not constitute a new result. Further, when spin is taken into account, it has been reported that a small gap does appear for zigzag nanoribbons[16]. Above discussion only pertains to uniform axial deformations, for ribbons with unit cell as indicated in Figure 1. Nonuniform deformations for which $t^i_{12} \neq t^i_{12N}$ can result in creation of band gap in zigzag nanoribbon. Such deformations can for example be sinusoidal deformations along the ribbon direction.

While we found that zigzag nanoribbons obtain no band gap due to stretching, armchair GNRs, on the other hand, can obtain nonzero band gap with the appropriate choice of hopping parameters. Hamiltonian matrix for armchair nanoribbons is given in Appendix. Fig. $1(b)$, shows the unit cell and hopping parameters for armchair nanoribbon. In general, under different deformations hopping parameters are not the same for different atoms in the unit cell. For example t^1_{12} may not equal t^1_{23} for different stretching deformations (see Fig. 2). To illustrate this idea, consider the situation when $t^i_{12} = -3.033eV$ and $t^i_{23} = t^i_{41} = -1.5eV$. Fig. 2 shows band gap due to changes in hopping parameter.

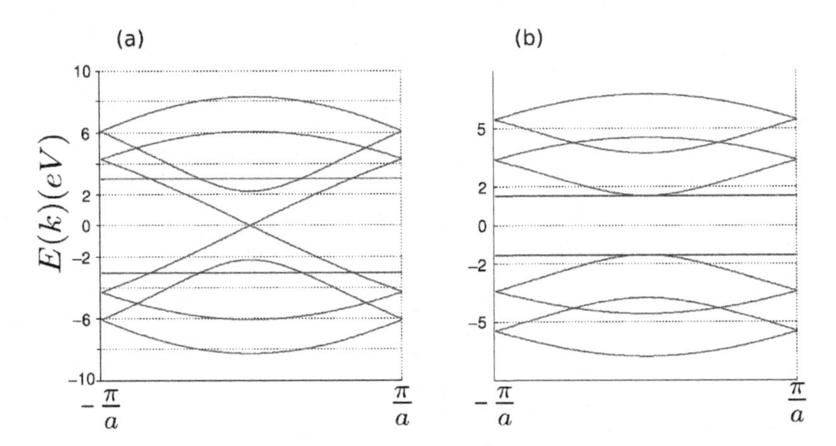

Fig. 2. Band gap variation with changes in hopping parameters for armchair ribbons, (*a*) all hopping parameters are the same, (*b*) hopping terms, $t^i_{12} = t^i_{34} = -3.033eV$, $t^i_{23} = t^i_{41} = -1.5eV$.

Our main objective of this paper is to consider twisting of graphene nanoribbons and changes of electronic structure to nonuniform deformations. Based on the simple Hückel analysis above, we expect zigzag nanoribbons to develop no band gap variation due to twist, since in that case condition (4) still holds. In case of armchair nanoribbons, there is no reason to expect zero band gap, since as we have seen, variation of hopping parameters can affect the band gap structure of armchair nanoribbons. In order to investigate it in some detail we use Density Functional Theory Tight-Binding Approximation (DFT-TBA) formulation[17,18].

3. Tight-Binding Approximation

To prepare twisted nanoribbons for TB calculations we start by constructing twisted nanoribbon using an algebraic formula, and later allow relaxation of twisted ribbons in order to reduce the total energy. Typically a ribbon relaxes into appropriate twisted configuration which correspond to local energy minima of the structure. Fig. 3 (*a*) shows a typical twisted armchair ribbon, while the corresponding band gap structure is shown in (*b*). Here the periodicity of the twist is 10.226 nm. Considering electronic structure of twisted ribbon we find zero band gap. Analyzing Fig. 3 we see that the inter-atomic bonds are adjusted in order to minimize structural energy. This adjustment leads to proper changes in hopping parameters such that no band gap is produced due to twisting deformation.

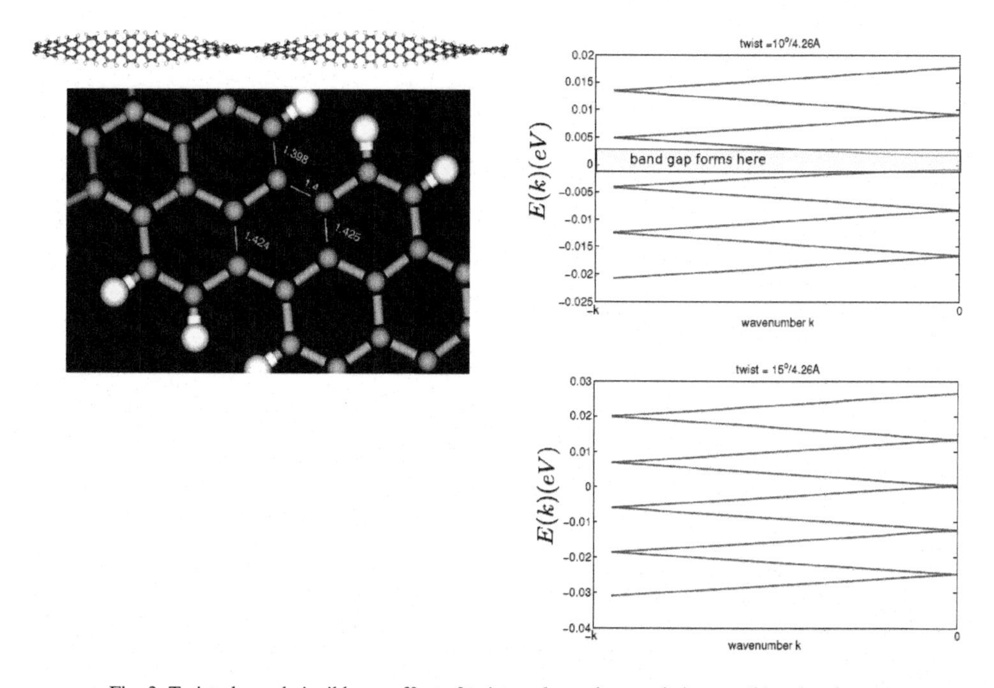

Fig. 3. Twisted armchair ribbons, effect of twist angle on the armchair nanoribbon band gap.

4. Inhomogeneous Deformations

While homogeneous twists do not produce band gap opening in armchair and zigzag ribbons, inhomogeneous deformations have potential for a band gap opening. To investigate band gap creation in zigzag and armchair nanoribbons we considered sinusoidal deformations of various amplitude but the same wavelength. We chose to investigate a nanoribbon unit cell of 42.6 Angstrom long and 4 zigzag layers width ($N = 5$, for armchair chains). Fig. 4 shows the typical deformation of nanoribbon. We have altered amplitude of the oscillations and are presenting band gap modulation for amplitude $A = 4$ and 8 Angstrom in Fig. 5 and 6 below.

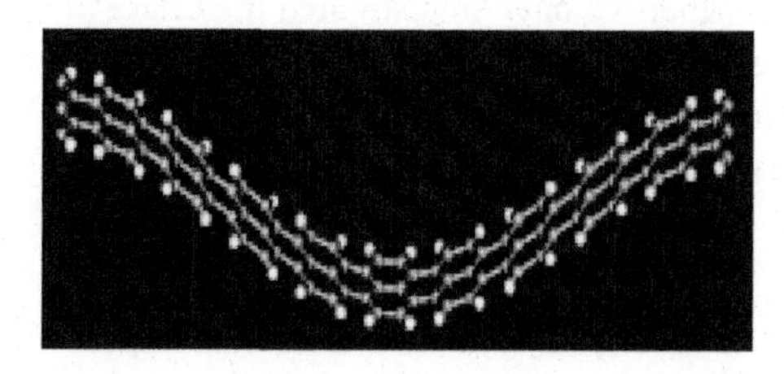

Fig. 4. Waved armchair nanoribbon – typical deformation.

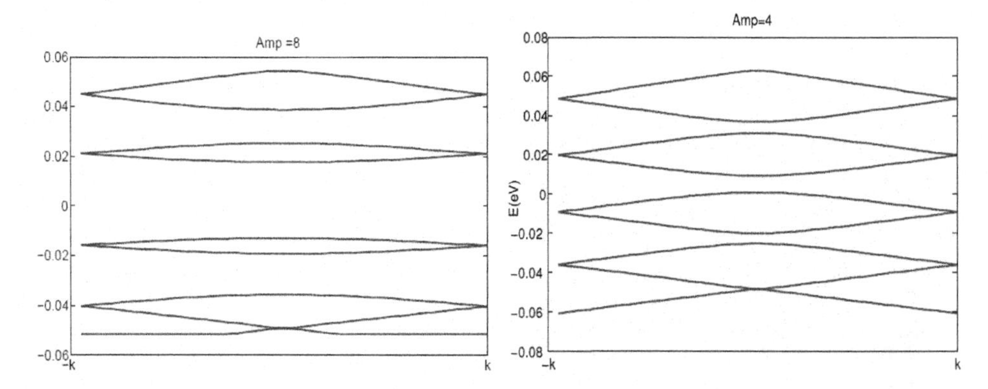

Fig. 5. Inhomogeneous deformations in armchair nanoribbons.

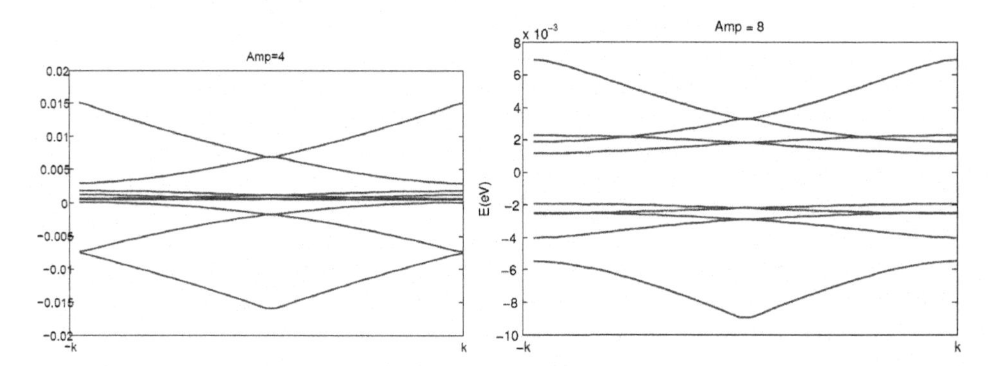

Fig. 6. Inhomogeneous deformations in zigzag nanoribbons.

5. Conclusion

We found that graphene nanoribbons exhibit band gap modulation when subjected to twisting and nonuniform deformations. We find that zigzag nanoribbons show no band gap changes due to twist, while twist deformations generate band gap opening in armchair nanoribbons. Further we have investigated the effect of waviness on nanoribbon electronic structure. We find that for both types of ribbons band gap is opened when the induced wave amplitude is at about 8 Angstrom. We thank the WCU program at SNU for its support of this research.

Appendix A.

$$H = \begin{bmatrix} \alpha_1 & \beta_1 & \cdots & 0 & 0 \\ \tilde{\beta}_2^T & \alpha_2 & \beta_2 & 0 & 0 \\ \vdots & \ddots & \ddots & \ddots & \\ 0 & 0 & \tilde{\beta}_m^T & \alpha_m & \gamma \\ 0 & 0 & 0 & \tilde{\gamma}^T & 0 \end{bmatrix} \tag{A.1}$$

$$\alpha_i = \begin{pmatrix} 0 & t_{12}^i & 0 & t_{14N}^i e^{-ika} \\ \tilde{t}_{12}^i & 0 & t_{23}^i & 0 \\ 0 & \tilde{t}_{23}^i & 0 & t_{34}^i \\ t_{14N}^i e^{ika} & 0 & \tilde{t}_{34}^i & 0 \end{pmatrix} \tag{A.2}$$

$$\beta_i = \begin{pmatrix} 0 & 0 & 0 & 0 \\ \tau_{12}^i & 0 & 0 & 0 \\ 0 & 0 & 0 & \tau_{34}^i \\ 0 & 0 & 0 & 0 \end{pmatrix} \tag{A.3}$$

$$\gamma = \begin{pmatrix} 0 & 0 \\ h_{12} & 0 \\ 0 & h_{32} \\ 0 & 0 \end{pmatrix} \tag{A.4}$$

Here t_{lm}^i, τ_{lm}^i and h_{lm} are different hopping parameters. All the hopping parameters are illustrates in Fig. 1.

References

1. Novoselov, A.H., Castro, N.F., Guinea, N.M.R., Peres, K.S. & Geim, A.K. The electronic properties of graphene. *Rev. Mod. Phys.* **81**, 110 (2009).
2. Geim, A.K. & Novoselov, K.S. The rise of graphene. *Nature Materials* **6**, 183-191 (2007).
3. Obradovic, B., *et al.* Analysis of graphene nanoribbons as a channel material for field-effect transistors. *Applied Physics Letters* **88**, 142102 (2006).
4. Han, M.Y., Ozyilmaz, B., Zhang, Y. & Kim, P. Energy Band-Gap Engineering of Graphene Nanoribbons. *Phys. Rev. Lett.* **98**, 206805 (2007).
5. McCann, E. Asymmetry gap in the electronic band structure of bilayer graphene. *Phys. Rev. B* **74**, 161403 (2006).
6. Ritter, C., Makler, S.S. & Latge, A. Energy-gap modulations of graphene ribbons under external felds: A theoretical study. *Phys. Rev. B* **77**, 195443 (2008).
7. Son, Y., Cohen, M.L. & Louie, S.G. Half-metallic graphene nanoribbons. *Nature* **444**, 347-349 (2006).
8. Linb, C.P., Changa, B.R., Wu, R.B. & Chen, M.F. Deformation effect on electronic and optical properties of nanographite ribbons. *J. Appl. Phys.* **101**, 063506 (2007).
9. Sun, L., *et al.* Strain effect on electronic structures of graphene nanoribbons: A frst-principles study. *J. Chem. Phys.* **129**, 074704 (2008).
10. Yasumasa Hasegawa, Rikio Konno, Nakano, H. & Kohmoto, M. Zero modes of tight-binding electrons on the honeycomb lattice. *Phys. Rev. B* **74**, 033413 (2006).
11. Pereira, V.M., Castro Neto, A.H. & Peres, N.M.R. Tight-binding approach to uniaxial strain in graphene. *Phys. Rev. B* **80**, 045401 (2009).
12. Bets, K.V. & Yakobson, B.I. Spontaneous Twist and Intrinsic Instabilities of Pristine Graphene Nanoribbons. *Nano Res* **2**, 78-83 (2009).
13. Shenoy, V.B., Reddy, C.D., Ramasubramaniam, A. & Zhang, Y.W. Edge-Stress-Induced Warping of Graphene Sheets and Nanoribbons. *Phys. Rev. Lett.* **101**, 245501 (2008).
14. Wehling, T.O., Balatsky, A.V., Tsvelik, A.M., Katsnelson, M.I. & Lichtenstein, A.I. Midgap states in corrugated graphene: Ab initio calculations and effective field theory. *EPL (Europhysics Letters)* **84**, 17003 (2008).
15. Nakada, K., Fujita, M., Dresselhaus, G. & Dresselhaus, M.S. Edge state in graphene ribbons: Nanometer size effect and edge shape dependence. *Phys. Rev. B* **54**, 17954-17961 (1996).
16. Lu, Y. & Guo, J. Band Gap of Strained Graphene Nanoribbons. *arXiv:0912.2702v2 [cond-mat.mes-hall]* (2010).
17. Goringe, C.M., Bowler, D.R. & Hernandez, E. Tight-binding modelling of materials. *Reports on Progress in Physics* **60**, 1447-1512 (1997).
18. Porezag, D., Frauenheim, T., Kohler, T., Seifert, G. & Kaschner, R. Construction of tight-binding-like potentials on the basis of density-functional theory: Application to carbon. *Phys. Rev. B* **51**, 12947-12957 (1995).

LOW-FREQUENCY ELECTRONIC NOISE IN GRAPHENE TRANSISTORS: COMPARISON WITH CARBON NANOTUBES

GUANXIONG LIU

Nano-Device Laboratory, Department of Electrical Engineering and Materials Science and Engineering Program, Bourns College of Engineering, University of California – Riverside, Riverside, California 92521 USA
guliu@ee.ucr.edu

WILLIAM STILLMAN

Center for Integrated Electronics and Department of Electrical, Computer and Systems Engineering, Rensselaer Polytechnic Institute, Troy, New York 12180 USA
stillw2@rpi.edu

SERGEY RUMYANTSEV

Center for Integrated Electronics and Department of Electrical, Computer and Systems Engineering, Rensselaer Polytechnic Institute, Troy, New York 12180 USA

Ioffe Physico-Technical Institute, Russian Academy of Sciences, St. Petersburg, 194021 Russia
roumis2@rpi.edu

MICHAEL SHUR

Center for Integrated Electronics and Department of Electrical, Computer and Systems Engineering, Rensselaer Polytechnic Institute, Troy, New York 12180 USA
shurm13@gmail.com

ALEXANDER A. BALANDIN*

Nano-Device Laboratory, Department of Electrical Engineering and Materials Science and Engineering Program, Bourns College of Engineering, University of California – Riverside, Riverside, California 92521 USA
balandin@ee.ucr.edu

**Corresponding author: email: balandin@ee.ucr.edu, web: http://ndl.ee.ucr.edu*

We report results of the experimental investigation of the low-frequency noise in graphene transistors. The graphene devices were measured in three-terminal configuration. The measurements revealed low flicker noise levels with the normalized noise spectral density close to *1/f* (*f* is the frequency) and the Hooge parameter $\alpha_H \sim 10^{-3}$. Both top-gate and back-gate devices were studied. The analysis of the noise spectral-density dependence on the gate biases helped us to elucidate the noise sources in these devices. We compared the noise performance of graphene devices with that of carbon nanotube devices. It was determined that graphene devices works better than carbon nanotube devices in terms of the low-frequency noise. The obtained results are important for graphene electronic, communication and sensor applications.

Keywords: graphene; 1/f noise; carbon nanotubes; low-frequency noise; graphene transistors.

1. Introduction

Superior properties of graphene [1-5] such as its extremely high room temperature electron mobility [1-3] and thermal conductivity [4-5] make this material appealing for electronic and sensor applications. Downscaling of modern electronic devices often results in an increasing noise level. Most of the proposed graphene applications require very low levels of the electronic flicker noise, which dominates the noise spectrum at low frequencies $f < 100$ kHz. The flicker noise spectral density is proportional to $1/f^{\gamma}$, where γ is a constant close to 1. The up-conversion of noise, which is unavoidable in electronic systems, results in serious limitations for practical applications. However, carbon materials, with strong covalent bonds, are less subject to electromigration or defect propagations [22], and, as a result, may likely to be a good material choice for low-noisy devices. Thus, it is important to investigate the noise level in graphene devices and identify its sources.

Two device configurations are studied in this research: conventional back-gate graphene devices and top-gated devices. The back gates in these devices are separated from the graphene channel by 300 nm of SiO_2 required for graphene optical visualization [1-3]. The top-gate device utilizes an additional layer of the gate oxide on top of the graphene. The top-gate configuration often leads to mobility degradation and may increase the noise. At the same time, the top gates are more practical and allow for a more detail study of the noise sources. In this paper, we report our results of the experimental investigation of the low-frequency noise in the top-gate and back-gate graphene field-effect transistors. For our studies, we selected devices with single layer graphene (SLG) and bi-layer graphene (BLG) channels and used HfO_2 as the top-gate dielectric.

2. Device Fabrication

We prepared graphene flakes with the lateral sizes of ~10 μm by mechanical exfoliation from the bulk highly oriented pyrolytic graphite (HOPG). The number of graphene layers and their quality were verified using the micro-Raman spectroscopy via the deconvolution of the Raman *2D* band and comparison of the intensities of the *G* peak and *2D* band [9-11]. The electron beam lithography (EBL) was used to define the regions for the source and drain electrodes, and a conductive Si substrate acted as the back gate (see Figure 1). For the top gate device, the gate oxide on the graphene flake was also defined by EBL and was followed by the low temperature atomic layer deposition (ALD). The thickness of HfO_2 directly deposited on top of the graphene channel was ~22 nm. A second step of EBL defined the source, drain and the top gate, and was followed by the electron beam evaporation to make Ti/Au electrodes. This sequence helped us to avoid possible damage to the contacts during to the long ALD process in the presence of H_2O and precursor environment. Figure 2 shows schematic of the top-gate graphene transistor structure and an optical microscopy image of a typical device (yellow colour corresponds to the metal contacts; green to HfO_2 dielectric and brown to SiO_2 dielectric).

The DC characteristics were measured both at UCR and RPI using a semiconductor parameter analyzer (Agilent 4156B). The examples of top-gate and back-gate current-voltage characteristics are shown in Figures 3 and 4. The charge neutrality point (the Dirac voltage) under the top-gate portion of the graphene transistor channel was V_{NP} = -1V (note the difference with that obtained by tuning the back-gate). The channel conductance was approximately proportional to the gate biases in both cases. The gate leakage current in the examined transistors was very small (~1 nA) indicating the quality of the gate dielectric. For the flakes with not perfect rectangular shape, the flake width was calculated by averaging the width along the channel length. The carrier mobility for our double-gate transistor with the carriers induced by the back gate was $\mu \approx 1550$ cm^2/Vs for electrons and $\mu \approx 2220$ cm^2/Vs for holes at room temperature. It was extracted with the help of the Drude formula, conventionally used for graphene devices. It give mobility μ as $\mu = g_m L/(V_D W C_{BG})$ [1-3], where g_m is the transconductance, L and W are the length and width of the channel, respectively, V_D is the drain voltage and C_{BG} is the capacitance of the substrate. The mobility for the top-gate transistors is 2~3 times lower than for our back-gate transistors. A possible reason for the reduction is related to extra scattering centers in the top gate dielectric, which scatter the electrons in graphene channel. We note that the Dirac points determined for the top-gate and back gate are not the same. The latter can be explained by differences in the surface states on the top and back gate oxides, which define the remaining charges in the channel. For this reason, different voltage is needed to compensate the changes and achieve the charge neutrality point.

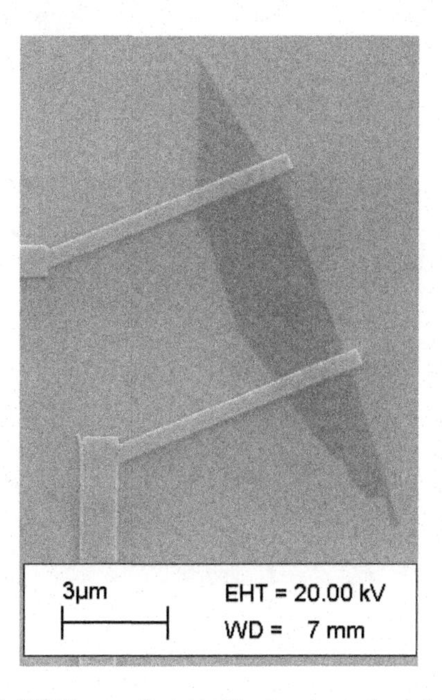

Fig. 1. SEM image of a typical back-gate graphene device.

3. Noise Measurements

The low-frequency noise was measured with a spectrum analyzer (SRS 770 FFT). The device bias was applied with a "quiet" battery potentiometer circuit in order to eliminate the possible noise generated by the power supplier. The drain bias was limited to 50 mV. The normalized current noise density S_I/I^2 for the top-gate graphene transistor is presented in Figure 5. As one can see, S_I/I^2 is very close to $1/f$ for the frequency f up to 3 kHz. The $1/f$ noise in electronic devices is often characterized by the empirical Hooge parameter

$$\alpha_H = \frac{S_I}{I^2} fN . \tag{1}$$

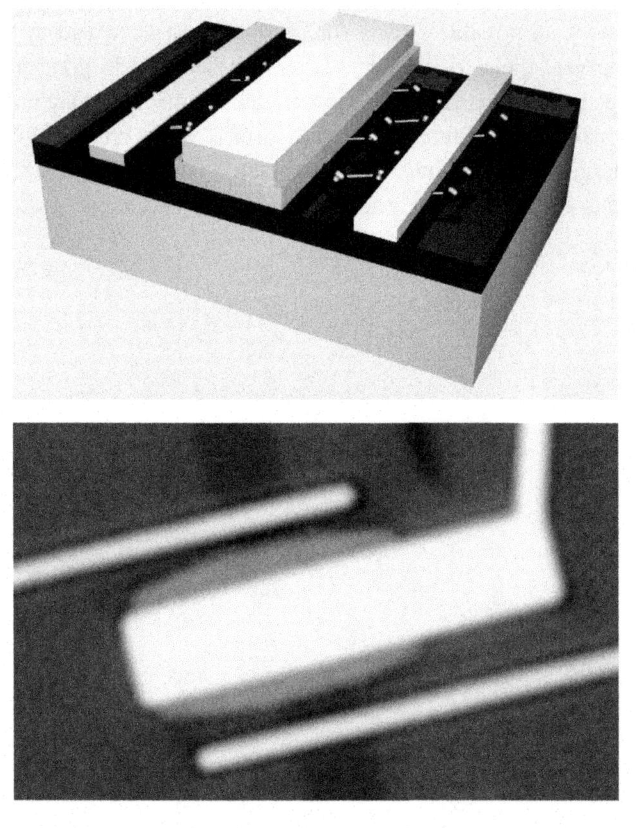

Fig. 2. Schematic of the top-gate graphene transistor (top panel) and an optical image of a typical graphene transistor (brown is SiO_2, yellow are metal gates and green is HfO_2).

Here N is the number of carriers in the channel estimated as $N = L^2/Rq\mu$ ($L \approx 9$ μm is the source – drain distance, R is the resistance from I-V measurements and q is the elemental charge). Using the measured mobility, we obtained the Hooge parameter $\alpha_H \approx 10^{-3}$. Such values are typical for many metals, semiconductor materials and devices [12]. In this sense, graphene transistors reveal similar noise characteristics as conventional devices. The extracted Hooge parameter can be used only for a rough comparison with other materials. In our analysis it does not imply a certain noise mechanism.

Figure 5 shows the gate voltage dependence of the noise S_I/I^2 at different frequencies. As seen, noise increases with the deviation of the gate voltage from the Dirac point, i.e. with increase of the electron or hole concentration in the graphene. Usually, good quality MOSFETs demonstrate the opposite tend, which is the decrease of the noise S_I/I^2 with the increase of the carrier concentration. This kind of dependence usually agrees very well (at least for n-channel MOSFETs) with McWhorter model which predicts $S_I/I^2 \sim 1/n_s^2$ (n_s is the electron channel concentration). Deviations from this law are often attributed to the influence of the contacts. Particularly, if the contact noise dominates, the spectral noise density can be presented in the form:

$$\frac{S_{Id}}{I_d^2} = \frac{S_C}{R_C^2} \frac{R_C^2}{(R_{ch} + R_C)^2}. \qquad (2)$$

where S_c/R_c is the spectral noise density of the contact resistance fluctuations, R_{Ch} and R_c are the channel and contact resistances, respectively. Equation (2) indicates that noise should increase with the decrease of the channel resistance provided $R_{Ch} > R_c$ and S_c/R_c does not depend on the gate voltage.

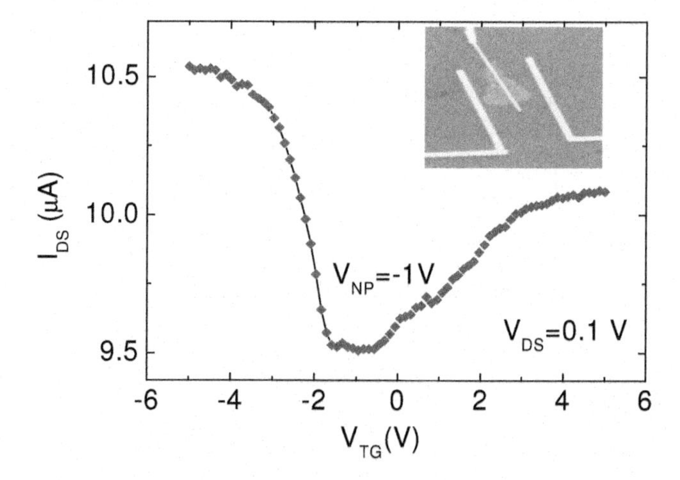

Fig. 3. Source-drain current as a function of the top gate demonstrating the top-gate action for the back-gate bias $V_{BG} = 0$ V. The inset shows an optical image of the measured transistor. The blue colour stripe under the top electrodes is graphene while the green region is HfO_2.

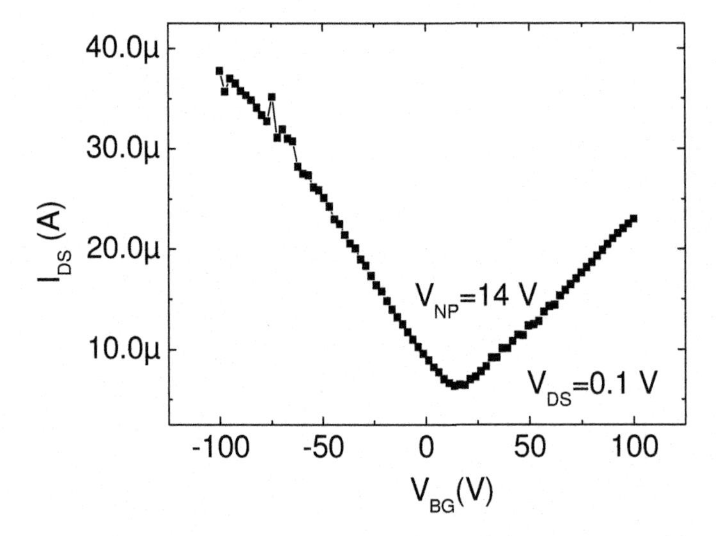

Fig. 4. Source-drain current as a function of the back gate bias demonstrating the back-gate action at the top-gate bias of $V_{TG} = 0$ V for the same HfO_2-graphene-SiO_2 double-gate transistor.

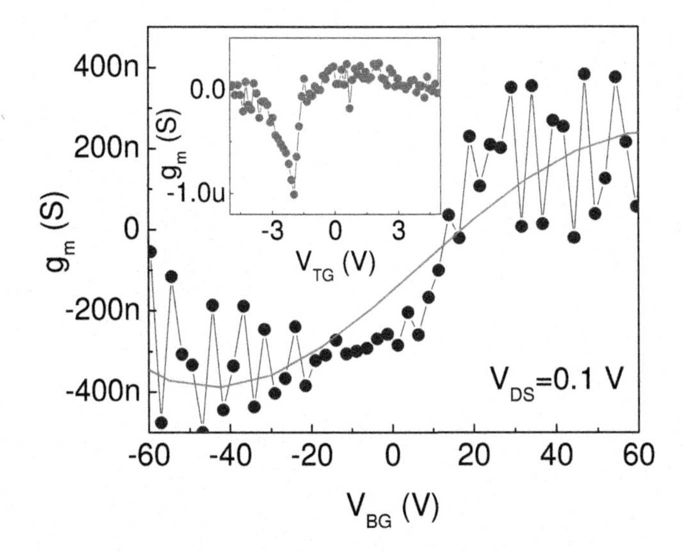

Fig. 5. Transconductance as a function of the gate voltage at the source-drain bias of 0.1 V. The inset shows tha data for the top-gate. The red curve is the fitting for the transconductance as a function of the back-gate bias.

Our measurements of the graphene layers with several contacts and analysis of the back-gate graphene transistors current-voltage characteristics showed that even at very high electron (hole) concentration the contact resistance is still smaller than the channel resistance for the majority of devices. The high contact noise is one of the possible explanations for the noise increase with the decreasing channel resistance.

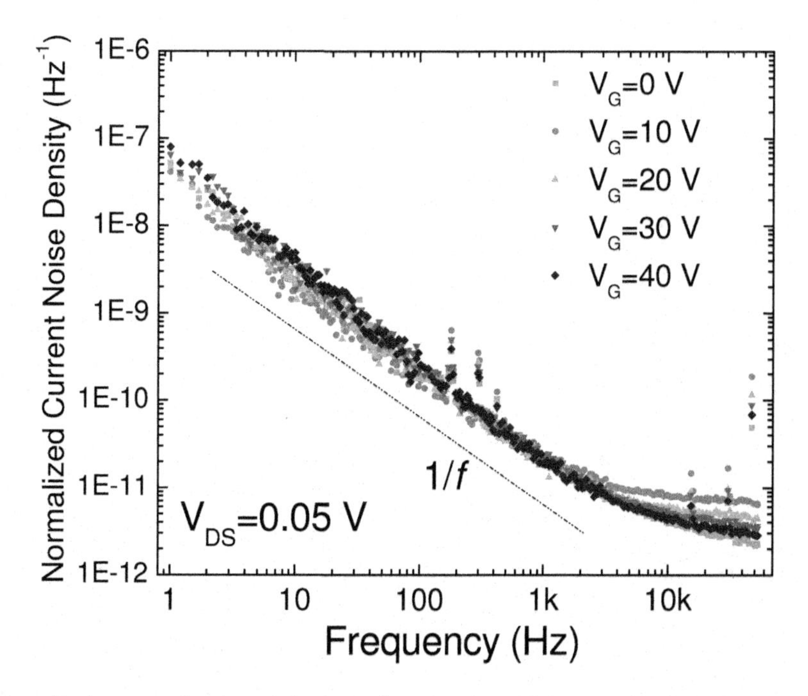

Fig. 6. Normalized current noise spectral density S_I/I^2 as a function of frequency f for the double-gate graphene transistor. The data is shown for the back-gate bias in the range from zero to 40 V. The *1/f* spectrum is indicated with the dashed line.

We did not observe any clear signatures of the generation – recombination noise (G-R) in the double-gate graphene transistors. This indicates the absence of well-defined traps contributing to noise.

4. Comparison with Carbon Nanotubes

The properties of carbon nanotubes (rolled up graphene) including the low frequency noise have been extensively studied in recent years. In Ref. [19] an empirical formula for noise in nanotubes was proposed:

$$\frac{S_I}{I^2} = \frac{A}{f} = \frac{10^{-11} R_n}{f}. \tag{3}$$

where R_n is the nanotube resistance. Even though, the experimental data for the nanotubes noise deviate from Eq. (3), parameter A is often plotted as a function of the nanotube resistance in order to compare the noise in different nanotube samples.

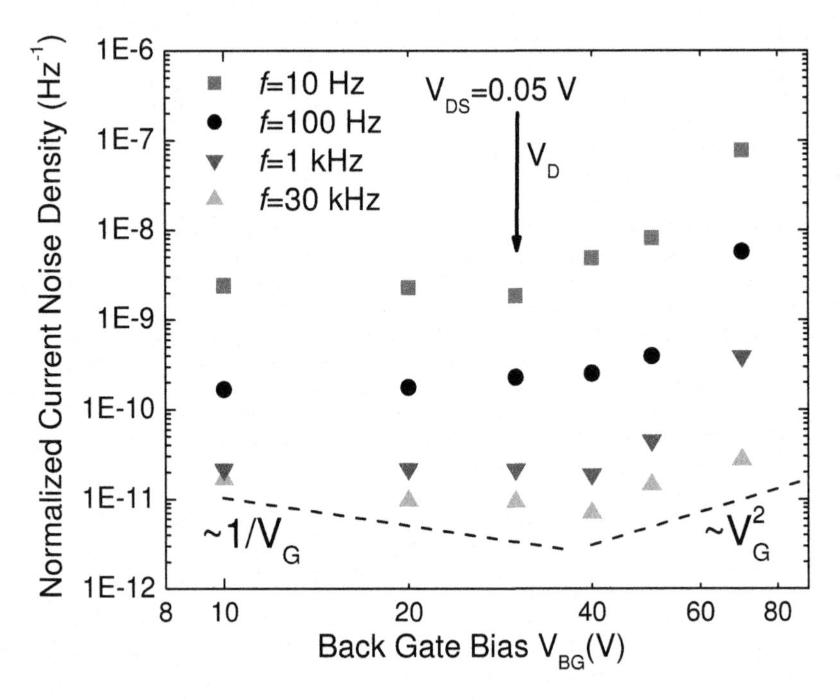

Fig. 7. Normalized current noise spectral density as a function of the back-gate bias for several frequencies. The drain – source bias was kept at 0.05 V.

Figure 8 compares noise of CNT from several groups with our graphene results (line corresponds to $A = 10^{-11}R$). As seen, the graphene devices are characterized by the lower noise amplitude than carbon nanotubes. However, this difference might be related to a much higher contact noise for carbon nanotubes (and higher contact resistance), since, in contrast to graphene transistors, the contact noise dominates in carbon nanotubes [18].

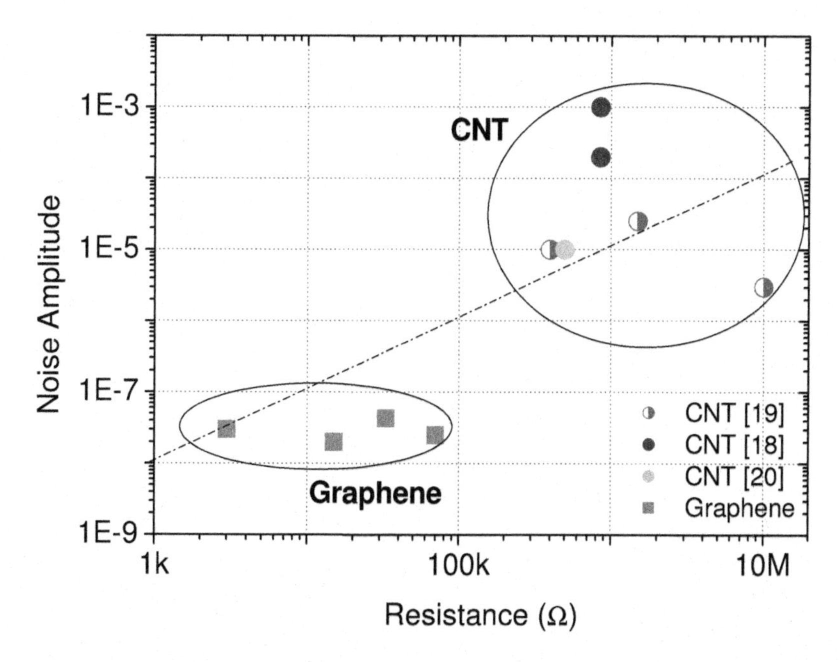

Fig. 8. Noise amplitude verses sample resistance of CNT and graphene. The noise data for CNTs are taken from literature [18-20]. The dashed line is an empirical dependence from Ref. [19].

5. Conclusions

We have studied experimentally *1/f* noise in the top-gate and back-gate graphene transistors. The *1/f* noise level in our graphene transistors with the bottom and top gates was rather low with the Hooge parameter $\alpha_H \approx 10^{-3}$. The spectrum noise density S_I/I^2 increased with the deviation of the gate voltage from the charge neutrality point. This dependence is opposite to that usually observed in semiconductor field effect transistors. Contribution of the contact noise to the overall transistor noise is one of the possible explanations for this dependence. The comparison with the carbon nanotubes showed that graphene transistors are characterized by the lower noise level. However, more detailed noise studies for carbon nanotubes are required to separate large contributions from the contact noise that might be responsible for this difference.

Acknowledgements

The work at UCR was supported by DARPA – SRC Focus Center Research Program (FCRP) through its Center on Functional Engineered Nano Architectonics (FENA). The work at RPI was supported by the IFC seed funding.

References

1. K. S. Novoselov, A. K. Geim, S. V. Morozov, D. Jiang, Y. Zhang, S. V. Dubonos and *et al*, *Science* **306**, 666 (2004); K. S. Novoselov, A. K. Geim, S. V. Morozov, D. Jiang, M. I. Katsnelson, I. V. Grigorieva and *et al*, *Nature* **438**, 197 (2005).
2. Y. B. Zhang, Y. W. Tan, H. L. Stormer and P. Kim, *Nature* **438**, 201 (2005); Y. W. Tan, Y. B. Zhang, H. L. Stormer and P. Kim, *Eur. Phys. J. Special Topics* **148**, 15 (2007).
3. S. V. Morozov, K. S. Novoselov, M. I. Katsnelson, F. Schedin, D. C. Elias, J. A. Jaszczak and *et al*, *Phys. Rev. Lett.* **100**, 016602 (2008).
4. A. A. Balandin, S. Ghosh, W. Bao, I. Calizo, D. Teweldebrhan, F. Miao and *et al*, *Nano Lett.* **8**, 902 (2008).
5. S. Ghosh, I. Calizo, D. Teweldebrhan, E. P. Pokatilov, D. L. Nika, A. A. Balandin and *et al*, *Appl. Phys. Lett.* **92**, 151911 (2008).
6. Y. M. Lin and P. Avouris, *Nano Lett.* **8**, 2119 (2008).
7. Q. Shao, G. Liu, D. Teweldebrhan, A. A. Balandin, S. Rumyantsev, M. Shur and D. Yan, *IEEE Electron Device Lett.* **30**, 288 (2009).
8. I. Meric, M. Y. Han, A. F. Young, B. Ozyilmaz, P. Kim and K. Shepard, *Nature Nanotechnology* **3**, 654 (2008).
9. I. Calizo, F. Miao, W. Bao, C. N. Lau and A. A. Balandin, *Appl. Phys. Lett.* **91**, 071913 (2007); I. Calizo, A. A. Balandin, W. Bao, F. Miao and C. N. Lau, *Nano Lett.* **7**, 2645 (2007).
10. I. Calizo, W. Bao, F. Miao, C. N. Lau and A. A. Balandin, *Appl. Phys. Lett.* **91**, 201904 (2007).
11. D. Teweldebrhan and A. A. Balandin, *Appl. Phys. Lett.* **94**, 013101 (2009).
12. A. A. Balandin, *Noise and Fluctuation Control in Electronic Devices* (American Scientific Publishers, Los Angeles, 2002).
13. J. M. Peransin, P. Vignaud, D. Rigaud and L. K. Vandamme, *IEEE Trans. Electron Devices* **37**, 2250 (1990).
14. A. A. Balandin, *Electron Lett.* **36**, 912 (2000).
15. V. Mitin and C. van Vliet, *Phys. Rev. B* **41**, 5332 (1990).
16. A. Svizhenko, S. Bandyopadhyay and M. A. Stroscio, *J. Phys.: Condensed Matter* **11**, 3697 (1999).
17. A. L. McWhorter, *Semiconductor Surface Physics* (University of Pennsylvania Press, Philadelphia, 1957).
18. A. Vijayaraghavan, S. Kar, S. Rumyantsev, A. Khanna, C. Soldano, N. Pala and *et al*, *J. Appl. Phys.* **100**, 024315 (2006).
19. P. G. Collins, M. S. Fuhrer and A. Zettl, *App. Phys. Lett.* **76**, 894 (2000).
20. S. K. Kim, Y. Xuan, P. D. Ye, S. Mohammadi, J. H. Back and M. Shim, *App. Phys. Lett.* **90**, 163108 (2007).
21. S. Russo, M. F. Craciun, M. Yamamoto, A. F. Morpurgo and S. Tarucha, arXiv:0901.0485v1 (2009).
22. B. Q. Wei, R. Vajtai and P. M. Ajayan, *App. Phys. Lett.* **79**, 1172 (2001).

ZnO NANOCRYSTALLINE HIGH PERFORMANCE THIN FILM TRANSISTORS

BURHAN BAYRAKTAROGLU, KEVIN LEEDY and ROBERT NEIDHARD

Air force Research Laboratory, AFRL/RYD
Wright Patterson Air Force Base, OH 45433, USA
burhan.bayraktaroglu@wpafb.af.mil

In this study, nc-ZnO films deposited in a Pulsed Laser Deposition (PLD) system at various temperatures were used to fabricate high performance transistors. As determined by Transmission Electron Microscope (TEM) images, nc-ZnO films deposited at a temperature range of 25°C to 400°C were made of closely packed nanocolums showing strong orientation. The influences of film growth temperature and post growth annealing on device performance were investigated. Various gate dielectric materials, including SiO_2, Al_2O_3, and HfO_2 were shown to be suitable for high performance device applications. Bottom-gate FETs fabricated on high resistivity (>2000 ohm-cm) Si substrates demonstrated record DC and high speed performance of any thin film transistors. Drain current on/off ratios better than 10^{12} and sub-threshold voltage swing values of less than 100mV/decade could be obtained. Devices with 2μm gate lengths produced exceptionally high current densities of >750mA/mm. Shorter gate length devices (L_G=1.2μm) had current and power gain cut-off frequencies, f_T and f_{max}, of 2.9GHz and 10GHz, respectively.

Keywords: Nanocrystalline; ZnO; thin films; pulsed lased deposition; atomic layer deposition; FET.

1. Introduction

Thin film transistors (TFT) made from amorphous or organic semiconductors are commonly used in the control circuit of large area display electronics such as flat panel TV screens. They can also be used in applications requiring flexible or non-planar surfaces where the use of regular single crystal electronics is problematic. The usefulness of TFTs, however, has not been extended to high performance applications due to the orders of magnitude lower electronic properties of thin films compared to their single crystal counterparts. Very low electron mobilities (0.1–1 $cm^2/V.s$) typically associated with conventional TFTs based on amorphous Si and organic semiconductors together with poor threshold voltage control issues prevented these technologies from advancing to more demanding applications.[1] Nanocrystalline ZnO (nc-ZnO) thin films offer a unique solution to improving TFT performance to levels comparable to single crystal semiconductors while maintaining their thin film properties.[2]

High performance TFTs were fabricated using a variety of metal-oxide semiconductors such as ZnO[3,4], gallium indium zinc oxide (GIZO)[5], $InGaO_3$[6] and zinc tin oxide.[7] As a binary compound, ZnO represents the simplest form of metal-oxide semiconductors that are being developed for thin film electronics applications. Its simpler structure makes it easier to control its composition during film deposition and therefore many different types of growth techniques have been successfully implemented for thin film transistor applications, including sputtering,[8,9] PLD,[2,10] ALD,[11,12,13] MOCVD,[14] and spin coating.[15] In addition to structural simplicity, ZnO also has some other interesting properties that make it particularly attractive for thin film applications. First, ZnO is a wide bandgap semiconductor (E_g=3.37eV), which makes it transparent to infrared and visible light,[16] enables it to support high electric fields, and maintain low leakage current even at high temperatures. Second, it maintains most of its intrinsic single crystalline electronic properties in thin films that contain a large number of crystalline defects.[17] These two properties taken together, i.e. defect tolerant operation of a wide bandgap semiconductor, allow ZnO to exploit superior electronic and optical benefits of wide bandgap semiconductors and the application diversity of thin film electronics.

ZnO TFTs are currently all NMOS type devices with n-type conduction channels. As with all MOS-type FETs, ZnO TFT performance depends strongly on the properties of the ZnO and gate insulator thin films as well as the interface states between these layers. Therefore, improvements in both the ZnO and the gate insulator films must be considered together to achieve higher performance. PLD-grown nc-ZnO TFTs on Plasma Enhanced Chemical Vapor Deposited (PECVD) SiO_2 gate insulators have shown drain current on/off ratios better than 10^{12}, electron mobility of 110 cm^2/V.s, and current density of higher than 400 mA/mm of gate width.[18] Despite the fact that a relatively low dielectric constant gate insulator was used, the maximum device transconductance was 80 mS/mm for 2μm gate length devices. Higher dielectric constant gate insulators such as Al_2O_3 and HfO_2 are attractive choices[19,20] for improving transconductance values.

In this paper, we have systematically investigated the influence of ZnO growth conditions on the device performance by examining the nanocrystalline structures of the films by transmission electron microscopy (TEM), atomic force microscopy (AFM), X-ray diffraction and various electrical measurements on the fabricated devices. Film growth temperatures in the range of 25°C to 400°C were investigated. The influence of gate insulator was also investigated using ALD-grown high dielectric constant films such as Al_2O_3 and HfO_2. Finally, it was shown that 1.2μm gate length nc-ZnO TFTs are capable of microwave amplification at frequencies as high as 10GHz.

2. Experiment

2.1. *Thin Film Preparation and Characterization*

2.1.1. *PLD-grown ZnO Thin Films*

ZnO films were deposited in a Neocera Pioneer 180 pulsed laser deposition system with a KrF excimer laser (Lambda Physik COMPex Pro 110, λ=248 nm, 10 ns pulse duration). The base pressure of the chamber was 4×10^{-8} Torr. ZnO films were deposited with a laser energy density of 2.6 J/cm^2, laser repetition rate of 30 Hz, deposition temperature of 25°C to 400°C, and oxygen partial pressure of 10 mTorr during the deposition. The target was a 50 mm diameter by 6 mm thick sintered ZnO ceramic disk (99.999%). Additional process parameters are discussed elsewhere.[21]

The ZnO crystal structure was determined by using a PANalytical X'Pert Pro MRD x-ray diffractometer. The film morphology was analyzed with an FEI DB235 scanning electron microscope (SEM) and a JEOL 4000EX transmission electron microscope (TEM) operating at 400 kV. Surface roughness was measured with a Veeco Dimension 3000 atomic force microscope (AFM) and analyzed with SPIP image processing software.

2.1.2. *PECVD and ALD-grown Gate Insulator Films*

In this study, three different gate insulators were used. SiO_2 films grown in PlasmaTherm 790 PECVD system at 250°C were used for most of the devices discussed here. We have also used higher dielectric constant insulators grown by ALD. Al_2O_3 and HfO_2 films were deposited at 250°C in a Cambridge Nanotech Fiji F200 ALD chamber. The Al_2O_3 depositions used trimethylaluminum and water as the aluminum and oxygen source, respectively. The HfO_2 depositions used tetrakis(dimethylamido)hafnium(IV) as the hafnium source and a remote radio frequency 300W O_2 plasma as the oxygen source. Argon was used as a precursor carrier gas and plasma purge gas. Nine point wafer maps of film thickness were measured with a Horiba Jobin Yvon UVISEL spectroscopic ellipsometer. Typical 20 nm films exhibited a 1σ thickness variation of $< 1\%$.

2.1.3. *TFT Fabrication and Characterization*

Devices for low frequency device characterization were fabricated on 200nm thick SiO_2-covered Si wafers. Higher resistivity substrates (>2000 ohm.cm) were used for high frequency device fabrication to minimize capacitive parasitics. A bottom-gate configuration was used for device fabrication with a Ni/Au (5 nm/120 nm) gate below the gate insulator. Devices for low frequency characterization had multiple gate fingers, whereas high frequency devices had only two gate fingers as shown in Figure 1.

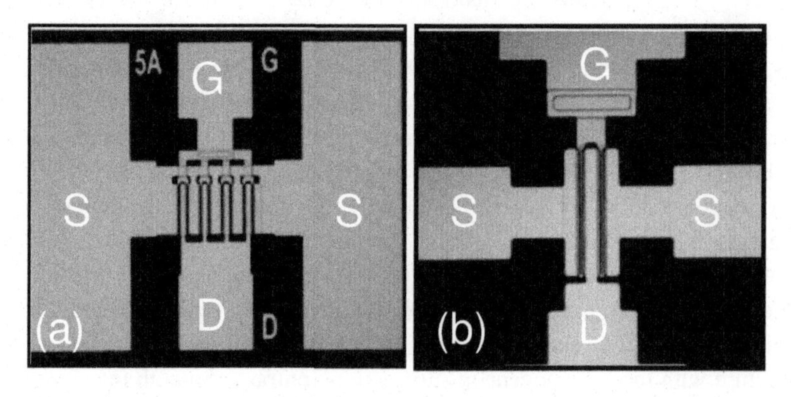

Figure 1: Images of nc-ZnO TFTs (a) Multi-finger device for low frequency characterization, W_G=400μm (b) High frequency device, L_G=1.2μm, W_G=100μm.

Identical device fabrication sequences were employed for the fabrication of low frequency and high frequency devices. The fabrication approach was typical of bottom-gate transistors and involved the fabrication of gate metal by evaporation and lift off techniques followed by the fabrication of the gate insulator and ZnO thin films over the gate metal. Based on the observation that the ZnO layer thickness in the range of 30-200 nm did not have a significant impact on the device performance; the film thickness was kept constant at 50 nm. As mentioned above, the ZnO deposition temperature ranged from 25°C to 400°C. Post-growth annealing was performed on some wafers in the temperature range of 400°C to 600°C in ceramic ovens containing clean room air. The device area was defined by mesa etching of the ZnO film in dilute HCl solution. Via holes were opened in the gate insulator layer over the gate contact pads using reactive ion etching before the fabrication of Ti/Pt/Au (20/30/350 nm) source/drain contacts by evaporation and lift of techniques. The source and drain contacts had various amounts of overlap to the gate electrode in three designs. The overlap amounts were approximately 1.5, 1, and 0.5μm, as measured by SEM, in designs designated as Design A, Design B, and Design C, respectively. No passivation layers were used over the top surface of the devices in this study.

The devices were dc characterized using an Agilent 4156C Precision Semiconductor Parameter Analyzer. Microwave performance was measured at room temperature using on-wafer coplanar microwave probes in an Agilent 8364B Precision Network Analyzer. From the measured *s*-parameters, current gain, $|h_{21}|^2$, and maximum available gain (MAG) values were determined as a function of frequency.

3. Results and Discussion

3.1. *Nanocrystalline ZnO Films*

X-ray diffraction scans of the ZnO films in Figure 2 exhibited a highly textured c-axis orientation with only the ZnO (002) peak present, consistent with other studies of PLD ZnO films.[22,23,24,25] The relative intensities of the (002) peak in films deposited between 25°C and 400°C generally increased with increasing deposition temperature. All 2Θ positions were less than the 34.421° 2Θ (002) peak from the JCPDS #36-1451 powder diffraction file indicating strained lattice structures. The full-width at half-maximum value for the (002) peak decreased with increasing deposition temperature indicating improved film crystallinity. Further details on XRD results are given elsewhere.[21]

Cross sectional TEM images reveal densely compacted, highly faulted columnar-shaped grains that predominantly extend through the thickness of the ZnO film as shown in Figure 3. At deposition temperatures of 75°C and below, diffraction contrast differences revealed elongated and, in and some areas, nearly equiaxed grain structures. Films deposited at 100°C and higher displayed a more uniform packed and elongated grain structure with more homogeneous diffraction contrast. Smooth interfaces between ZnO and SiO_2 are observed at all deposition temperatures while the surfaces of the ZnO grains are predominantly dome-shaped.

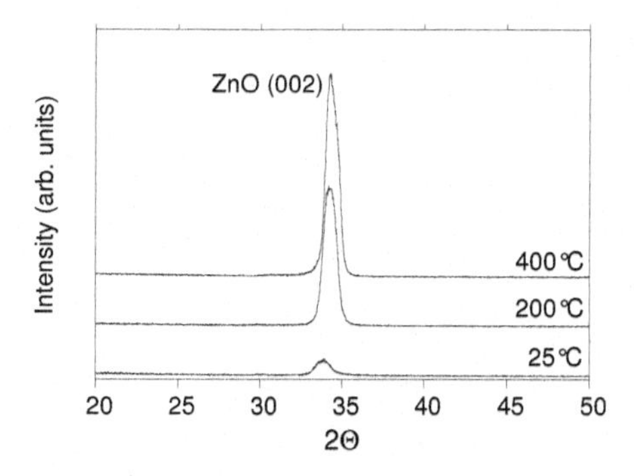

Figure 2: X-ray diffraction patterns of ZnO thin films deposited at 25°C to 400°C.

Corresponding surface SEM images in Figure 3 show uniform surface morphologies for all ZnO depositions with no trend in grain size as deposition temperature increases. AFM images with 500 nm x 500 nm collection areas in Figure 3 also show that the films were predominantly smooth. RMS roughness values ranged from 0.65 nm to 1.65 nm. Although the lowest roughness occurred in a film deposited at 200°C, no trend in roughness as a function of deposition temperature was observed. The RMS roughness is similar to other reported values of ZnO deposited by PLD.[24,25] Grain size calculations based on grain boundary intercepts in a scanning probe image processing software indicated 25 nm to 35 nm ZnO grains, again with no trend observed as a function of deposition temperature.

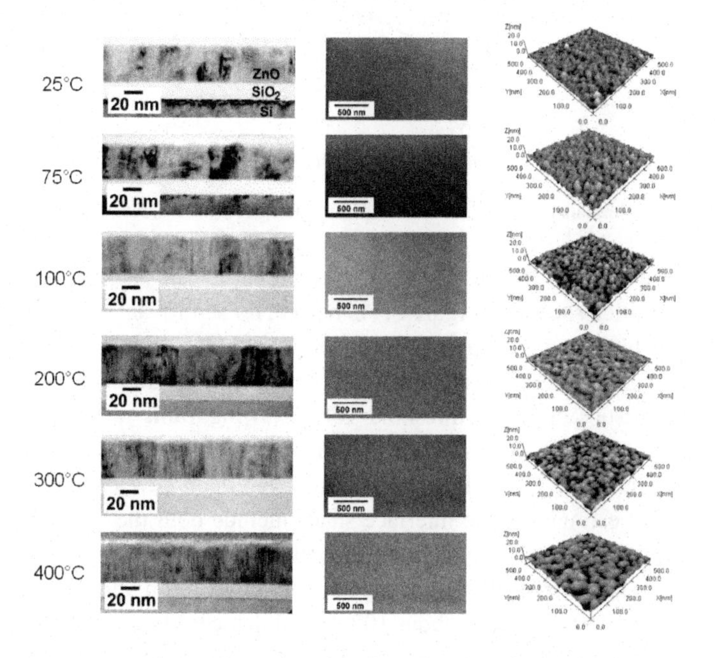

Figure 3: Cross sectional TEM images, surface SEM images and AFM images of ZnO thin film deposited on 20 nm SiO_2/Si at 25°C, 75°C, 100°C, 200°C, 300°C and 400°C.

3.2. *Low Frequency Devices*

The intrinsic properties of nc-ZnO TFTs were examined using various gate length devices (2 - 25μm). The common source I_D*L_G vs. V_D characteristics of devices with SiO_2 gate insulators and various gate lengths are shown in Figure 4. In the linear region, the devices all show nearly identical characteristics indicating normal device size scaling. In the saturated region, shorter gate length devices exhibit higher current levels due to channel length modulation (CLM)[26] with drain bias. This effect may be reduced by proportional decrease in t_{ox} or increase in channel doping density. The devices studied here all had t_{ox}=30nm and an estimated channel doping levels of $2x10^{16}$ cm^{-3}. Devices were fully depleted and the I-V characteristics did not exhibit any droop due to self heating. No significant hysteresis was observed with any gate length device.

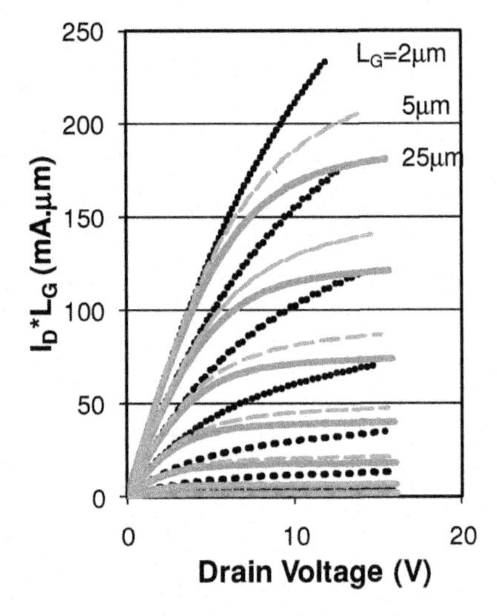

Figure 4: Normalized I-V characteristics of nc-ZnO TFTs with different gate lengths.

The influence of gate length on device performance can also be observed in the transfer characteristics shown in Figure 5. As before, no significant hysteresis effects are seen as the gate bias is swept in both directions. In the sub-threshold voltage region, where the drain current increases exponentially with gate bias, all devices exhibit almost identical characteristics. The sub-threshold voltage swing value in this region is about 75mV/decade, which corresponds to an interface state density of $2.5x10^{11}$ cm^{-2} using relative dielectric constant of 3.9 for SiO_2 and the expression developed in Ref. 27. Note that in this type of estimation, the interface states include both the surface and the bulk states and all such states are assumed to be independent of energy. A gate bias dependent electron mobility characteristics were observed with a maximum value of 110 cm^2/V.s.[18] The drain current on/off ratios are about 10^{12} for all devices. The parameter that is most impacted by the gate length is the drain current density, as expected from normal device size scaling.

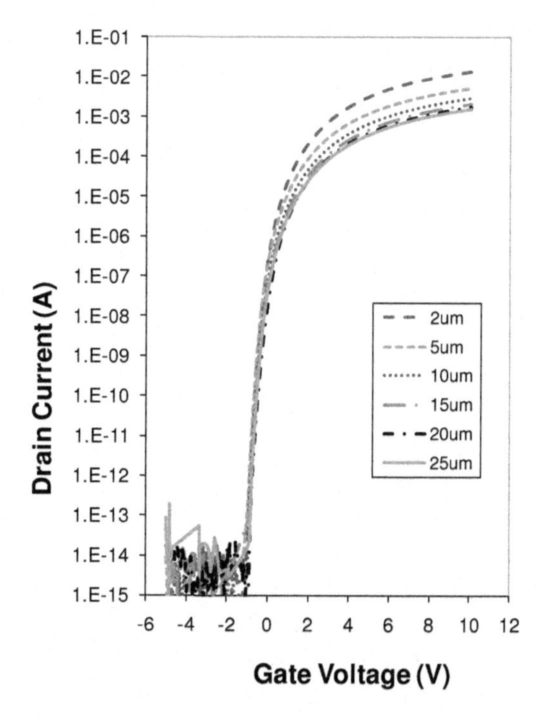

Figure 5: ZnO TFT transfer characteristics as a function of gate length. Drain current on/off ratios better than 10^{12} and sub-threshold voltage swing values of 75mV/decade were obtained with all size devices.

3.3. *Influence of ZnO Growth Temperature*

ZnO films grown at different temperatures all maintained their nanocolumnar structures, as discussed above. The electrical characteristics of the transistors made from these films also maintained their basic characteristics. All devices had excellent on/off ratios of about 10^{12} and no hysteresis characteristics. Threshold voltage values estimated from the linear portion of the I_D vs. V_G curves changed slightly from 1.05V to 0.9V as the growth temperature increased from 25°C to 400°C.

The only device parameter that seems to be influenced by the growth temperature is the drain current density, which is related to the film conductivity and the electron mobility. Figure 6 shows the current density of devices with different gate lengths whose ZnO films were grown at different temperatures. The current density increases with gate length linearly for devices grown at the same temperature as seen before (see Figure 4). All device sizes showed sharp increase in current density as the growth temperature increased from 25°C to 200°C. Beyond 200°C, a slight decrease in current density was observed. Beyond 400°C, the current density rapidly drops to near zero as nanocolumns begin to be separated from each other. Based on these observations, we speculate the packing of nanocolumns is maximized at 200°C.

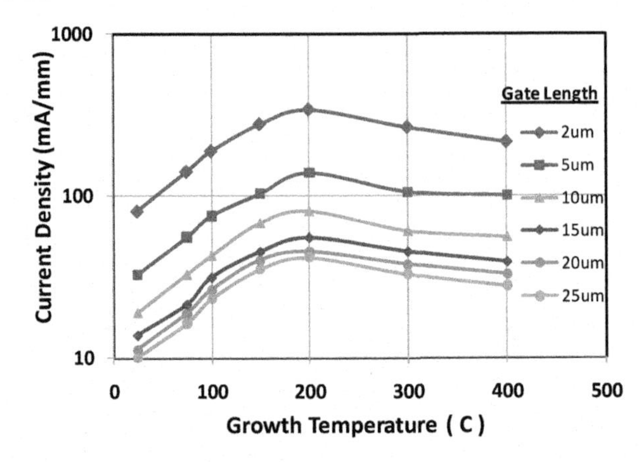

Figure 6: Current density dependence on film growth temperature and device gate length. Measurements were made on devices with WG=400μm at V_D=14V and V_G=12V.

Drain current density is further influenced by post-growth annealing of films prior to device fabrication. Films grown even at 400°C showed significant increase in current density after annealing at 400°C for 1hr. The current density increase was nearly 2X, 3X and 3.5X when annealed at 400°C, 500°C and 600°C, respectively, as shown in Figure 7. Based on the observations that the nanocrystal size increases with post-growth annealing, (to 40nm after 600°C anneal) the current density increase can be due to higher conduction across smaller number of grain boundaries for a given size device. From Figure 7, it can be estimated that devices with 1μm and 0.5μm gate lengths made with films post-growth annealed at 600°C will be capable of over 1.2A/mm and 2A/mm current density, respectively.

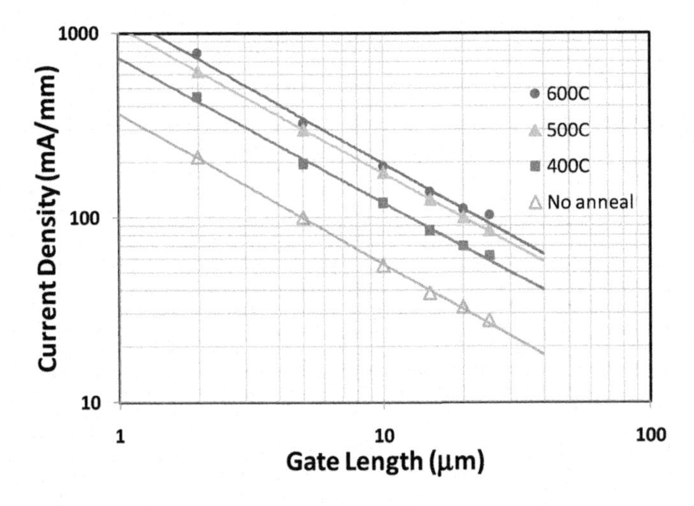

Figure 7: Current density dependence on post-growth annealing and device gate length. Measurements were made on devices with WG=400μm at V_D=14V and V_G=12V.

3.4. *Influence of Gate Dielectric*

Higher dielectric constant gate insulators can be used to enhance the field effect and increase the device transconductance. Short channel effects observed with low dielectric constant insulators such as SiO_2 (e.g. see Figure 4) can be mitigated using insulators with higher dielectric constants. We have used ALD-grown Al_2O_3 and HfO_2 in this work and analyzed the device performance enhancements with respect to the baseline SiO_2 gate insulators.

Figure 8 shows the I-V characteristics of three same size TFTs with 20nm thick gate insulators and 50nm thick ZnO layers grown at 200°C. The same gate bias was applied to all devices for a direct comparison. Because of differences in breakdown strengths in the insulators studied, the maximum gate bias was limited to 10V. This limitation was set by the breakdown voltage of HfO_2 at 5 x 10^6 V/cm. The breakdown strengths of the other insulators were about 2X higher.

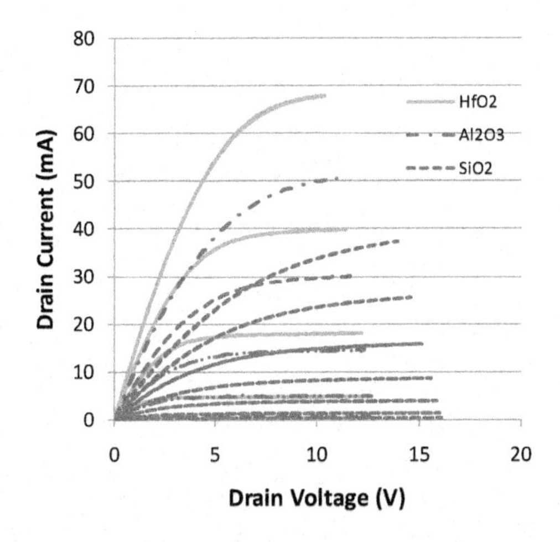

Figure 8: I-V characteristics of nc-ZnO TFTs with different gate insulators. The device size was L_G=5μm, W_G=400μm, t_{ox}=20nm. Gate bias was increased from 0 to 10V in 2 V/step increments.

It can be seen in this figure that all three insulators are suitable for ZnO TFT applications. The lack of hysteresis in device characteristics indicates the absence of bulk or interface traps. Higher and more saturated drain current characteristics are achieved with HfO_2 and Al_2O_3 than SiO_2 at the same gate bias values indicating that both the transconductance and short channel effects are improved. Figure 9 shows in more detail the improvements in device transconductance with the use of higher dielectric constant gate insulators. For all device sizes, from L_G=2μm to 25μm, the transconductance improved by a factor of 2 and 3.5 with respect to SiO_2 by the use of Al_2O_3 and HfO_2. This improvement factor is somewhat less than the ratios of dielectric constants. For example, the HfO_2/SiO2 dielectric constant ratio is about 5, whereas the transconductance improvement is only 3.5. The difference between these ratios can be explained by the presence of higher interface state density in HfO_2/ZnO than in SiO_2/ZnO. It is therefore possible that even higher transconductance values can be achieved with HfO_2 gate insulators with further improvements in interface quality. The highest transconductance value achieved was 135 mS/mm for 2μm gate length devices using HfO_2 gate insulator. This is substantially higher that the best value (80 mS/mm) previously obtained with SiO_2 gate insulators at higher gate bias conditions.[18]

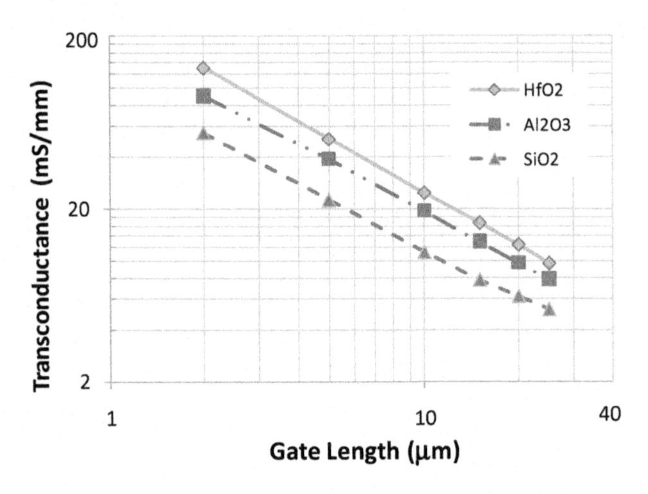

Figure 9: ZnO TFT tranconductance measured at V_D=12V and V_G=10V for 3 different gate insulators. L_G=5μm, W_G=400μm and t_{ox}=20nm.

3.5. *High Frequency Devices*

High frequency devices fabricated on high resistivity Si substrates were examined for their microwave signal amplification potential. The gate insulator for these devices was SiO_2 and the ZnO films were grown at 400°C. Three different device designs all with L_G=1.2μm and W_G=2x50μm, but with varying amounts of gate electrode overlap with source/drain contacts, were biased at V_G=6V and V_D=11V for small-signal microwave tests. From the measured s-parameters, the maximum available gain (MAG) and current gain ($|h_{21}|^2$) values were determined shown in Figure 10. All three device designs showed an identical current gain cut off frequency of 2.9GHz. The power gain cutoff frequencies of the same devices showed a strong dependence on the amount of gate contact overlap with source/drain contacts. Designs A, B, and C with overlap amounts of approximately 1.5, 1.0 and 0.5μm exhibited f_{max} values of 7.5GHz, 8.5GHz and 10GHz, respectively. The performance of Design A is similar to the previously reported value of 7.45GHz for the same size device.[28] The improved performances of Designs B and C indicate the importance of minimizing parasitic overlap capacitances. The results obtained here with nc-ZnO TFTs compare favorably with the RF performance obtained (f_T=180 MHz, f_{max}=155 MHz) with other metal oxide (InZnO) thin film transistors on glass substrates.[29] To the best knowledge of the authors, these are the highest cutoff frequencies obtained with thin film transistors, and indicate the excellent potential of nc-ZnO TFTs for RF circuit applications.

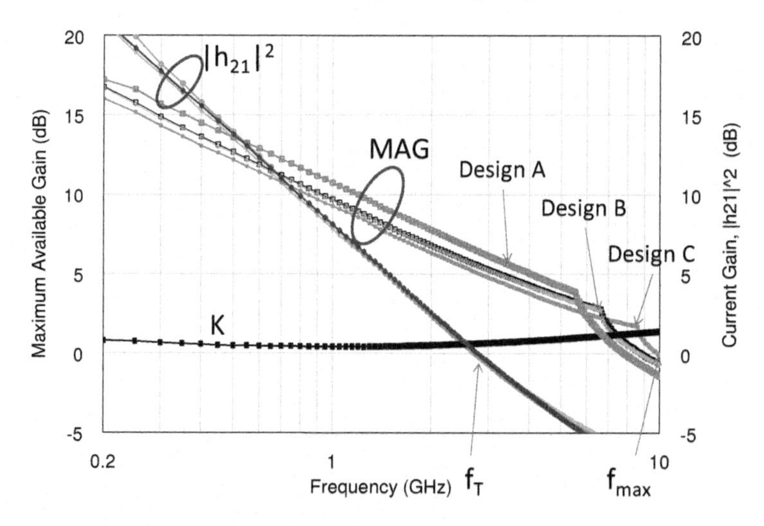

Figure 10: Small signal microwave characteristics of nc- ZnO TFTs.

4. Summary and Conclusions

It was shown that ZnO thin films prepared by the pulsed laser deposition technique across a wide temperature range (25°C to 400°C) have ordered nanocolumnar structures with column diameters of about 25-30nm. The diameters of columns increase to about 40nm with post-growth annealing. Most critical device parameters including drain current on/off ratios, sub-threshold voltage swings, threshold voltages and hysteresis-free operation were shown to be independent of film growth temperature. The only parameter that had dependence on growth temperature was the current density, which exhibited a peak for films grown at 200°C. The current density also showed a significant increase with post-growth annealing, possibly due to reduction of the number of grain boundaries with annealing. A substantial increase in device transconductance was obtained with the use of ALD-grown high dielectric gate insulators (Al_2O_3 and HfO_2) compared to SiO_2. A record transconductance of 135mS/mm was obtained with 2μm gate length devices using HfO_2 gate insulators. High frequency response of various device designs fabricated on Si substrates with 1.2μm gate lengths were shown to be sensitive to gate contact overlap with source and drain contacts. While the current gain cut off frequency was the same for all designs (f_T=2.9GHz), the power gain cut off frequency, f_{max}, varied from 7.5GHz to 10GHz depending on the amount of overlap.

Acknowledgements

This work was supported in part by Air Force Office of Scientific Research under LRIR 07SN03COR (Dr. K. Reinhart) and Defense Advanced Research Projects Agency (Dr. J. Albrecht). The authors thank D. Tomich and J. Brown for ZnO film characterization.

References

1. C. R. Kagan and P. Andry, *Thin Film Transistors* (Marcel Dekker Publishing, New York, 2003).
2. B. Bayraktaroglu and K. Leedy, *ECS Transactions* **16**, 61-73 (2008).
3. P. F. Carcia, R. S. McLean, and M. H. Reilly, *Appl. Phys. Lett.* **88**, 123509 (2006).
4. P. K. Shin, Y. Aya, T. Ikegami, and K. Ebihara, *Thin Solid Films* **516**, 3767 (2008).
5. W. Lim, S. H. Kim, Y. L. Wang, J. W. Lee, D. P. Norton, S. J. Pearton, F. Ren, and I. I. Kravchenko, *J. Electrochem. Soc.* **155**, H383 (2008).
6. H. Q. Chiang, D. Hong, C. M. Hung, R. E. Presley, and J. F. Wager, *J. Vac. Sci. Technol. B* **24**, 2702 (2006).
7. H. Q. Chiang, J. F. Wager, R. L. Hoffman, J. Jeong, and D. A. Keszler, *Appl. Phys. Lett.* **86**, 013503 (2005).
8. P. F. Carcia, R. S. McLean, M. H. Reilly, and G. Nunes, Jr., *Appl. Phys. Lett.* **82**, 1117-1119 (2003).
9. D. J. Kang *et.al.*, *Thin Solid Films* **475**, 160-165 (2005).
10. S. Masuda, K. Kitamura. Y. Okumura, S. Miyatake, H. Tabata, and T. Kawai, *J. App. Phys.* **93**, 1624-1630 (2003).
11. D. Mourey, D. A. Zhao, J. Sun, and T. N. Jackson, *IEEE Tran. Electron Dev.* **57**, 530 (2010).
12. W. S. Choi, *J. Soc. Inf. Display* **17**, 751-755 (2009).
13. K. Kopalko *et. al.*, *Phys. Stat. Sol. (c)* **2**, 1125-1130 (2005).
14. S. Yoshizawa, K. Nishimura, and T. Sakurai, *J. Phys: Conf. Series* **100**, 082052 (2008).
15. B. J. Norris, J. Anderson, J. F. Wager, and D. A. Keszler, *J. Phys. D: Appl. Phys.* **36**, L105-L107 (2003).
16. Ü. Özgür, Y. I. Alivov, C. Liu, A. Teke, M. A. Reshchikov, S. Doğan, V. Avrutin, S. J. Cho, and H. Morkoç, *Appl. Phys. Rev.* **98**, 041301 (2005).
17. K. Nomura, H. Ohta, A. Takagi, T. Kamiya, M. Hirano, and H. Hosono, *Nature* **432**, 488-492 (2004).
18. B. Bayraktaroglu, K. Leedy, and R. Neidhard, *IEEE Electron Dev. Lett.* **29**, 1024-1026 (2008).
19. N. C. Su, S. W. Wang, and A. Chin, *Electrochem. Solid-State Lett.* **13**, H8 (2010).
20. P. F. Carcia, R. S. McLean, and M. H. Reilly, *Appl. Phys. Lett.* **88**, 123509 (2006).
21. B. Bayraktaroglu, K. Leedy, and R. Neidhard, *Mater. Res. Soc. Symp. Proc.* **1201**, H09-07 (2010).
22. L. Bentes, R. Ayouchi, C. Santos, R. Schwarz, P. Sanguino, O. Conde, M. Peres, T. Monteiro, and O. Teodoro, *Superlattices and Microstructures* **42**, 152 (2007).
23. S. Amirhaghi, V. Craciun, D. Craciun, J. Elders, and I. W. Boyd, *Microelectronics Engineering* **25**, 321 (1994).
24. L. Han, F. Mei, C. Liu, C. Pedro, and E. Alves, *Physica E* **40**, 699 (2008).
25. C.-F. Yu, C.-W. Sung, S.-H. Chen, and S.-J. Sun, *Appl. Surface Sci.* **256**, 792 (2009).
26. J. J. Liou, A. Ortiz-Conde, and F. Garcia-Sanchez, *Analysis and Design of MOSFETs: Modeling, Simulation, and Parameter Extraction* (Kluwer Academic Publishers, Boston, 1998).
27. A. Roland, J. Richard, J. P. Kleider, and D. Mencaraglia, *J. Electrochem. Soc.* **140**, 3679 (1993).
28. B. Bayraktaroglu, K. Leedy, and R. Neidhard, *IEEE Electron Dev. Lett.* **30**, 946 (2009).
29. Y. L. Wang, L. N. Covert, T. J. Anderson, W. Lim, J. Lin, S. J. Pearton, D. P. Norton, J. M. Zavada, and F. Ren, *Electrochem. Sol. State Lett.* **11**, H60 (2008).

ZINC OXIDE NANOPARTICLES FOR ULTRAVIOLET PHOTODETECTION

SHAYLA SAWYER*, LIQIAO QIN and CHRISTOPHER SHING

Electrical, Computer, and Systems Engineering Department, Rensselaer Polytechnic Institute, 110 8th Street, Troy, NY 12180, United States
**sawyes@rpi.edu*

Zinc Oxide (ZnO) nanoparticles were created by a top-down wet-chemistry synthesis process (ZnO-A) and then coated with polyvinyl-alcohol (PVA) (ZnO-U). In ZnO-U, strong UV emission was apparent while the parasitic green emission, which normally appears in ZnO suspensions, was suppressed. A standard lift-off process via e-beam lithography was used to fabricate a detector by evaporating Aluminum (Al) as ohmic electrodes on the ZnO nanoparticle film. Photoconductivity experiments showed that linear current-voltage response were achieved and the ZnO-U nanoparticles based detector had a ratio of UV photo-generated current more than 5 times better than that of the ZnO-A based detector. In addition, non-linear current-voltage responses were observed when interdigitated finger Gold (Au) contacts were deposited on ZnO-U. The UV generated current to dark current ratios were between 4 and 7 orders of magnitude, showing better performance than the photodetector with Al contacts. ZnO-U were also deposited on Gallium Nitride (GaN) and Aluminum Gallium Nitride (AlGaN) substrates to create spectrally selective photodetectors. The responsivity of detector based on AlGaN is twice that of commercial UV enhanced Silicon photodiodes. These results confirmed that ZnO nanoparticles coating with PVA is a good material for small-signal, visible blind, and wavelength selective UV detection.

Keywords: ZnO; photodetectors; nanoparticle; ultraviolet; green photoluminescence; surface passivation.

1. Introduction

Zinc oxide (ZnO) is the emerging alternative semiconductor material to Gallium Nitride (GaN) for optoelectronic applications[1]. Its properties are close to that of GaN material with the additional advantages of lower cost, large-area native substrates, low temperature growth, and an exciton binding energy (60 meV) twice that of GaN (28 meV)[2-6]. The band gap of ZnO (~3.2-3.4 eV) is ideal for light-emitting diodes (LEDs), laser diodes, and photodetectors in the ultraviolet (UV) wavelength range. Its primary hindrance from greater prevalence is the lack of a stable and reproducible p-type material although current research with promising results have rejuvenated overall research interest[7].

Semiconductor photodetectors convert incident light within a range of wavelengths into detectable current based on their material properties. This wavelength range is tuned by the bandgap of the material. Currently, light detection in the blue/UV region is facilitated by Silicon (Si) photodetectors. At room temperature, the bandgap energy of Si (1.2 eV) is far below ideal for detection in the blue/UV region, which greatly reduces responsivity. Wider bandgap materials, including ZnO and AlGaN/GaN, are better matched in energy from the UV to blue wavelength region, resulting in an overall increase in photosensitivity to UV relative to Si. The efficiency and sensitivity of the conversion of incident radiation to current is dominated by limitations of area, light coupling, and wavelength specificity. Currently, many nanostructures of ZnO have been studied extensively due to their quantum confinement effects which correspond to continuous tuning of the spectral wavelength and improved device performance[8-16]. Low cost, large area, wavelength tunable, ultraviolet photodetectors are possible with the inherent properties of ZnO nanoparticles.

2. ZnO Background

The first studies of ZnO as a semiconductor material began in 1935 by C.W. Bunn with an investigation of lattice parameters. ZnO has three kinds of crystal structures, rocksalt, zinc blende and wurtzite, which are shown in Figure 1. In ambient conditions, the wurtzite structure is thermodynamically stable, so it is the most common structure of ZnO. Its hexagonal lattice is defined by two interconnected sublattices of Zn^{2+} and O^{2-}. The Zn^{2+} ion is surrounded by the O^{2-} ion and vice-versa, therefore, ZnO has polar surfaces with inherent piezoelectric and spontaneous polarization effects. The hexagonal lattice parameters are a = 3.25 Å, c = 5.2069 Å, and c/a = 1.60.

ZnO material is intrinsically n-type. While doping with donors, Aluminum, Gallium and Indium (Al, Ga, and In) up to 10^{20} cm^{-3} is readily achieved, acceptor doping has proven to be difficult[7]. Common acceptors include Nitrogen, Phosphorus, Arsenic and Antimony (N, P, As, and Sb) with up to 10^{15} cm^{-3} active dopants. This is the primary limiting factor for ZnO devices to date.

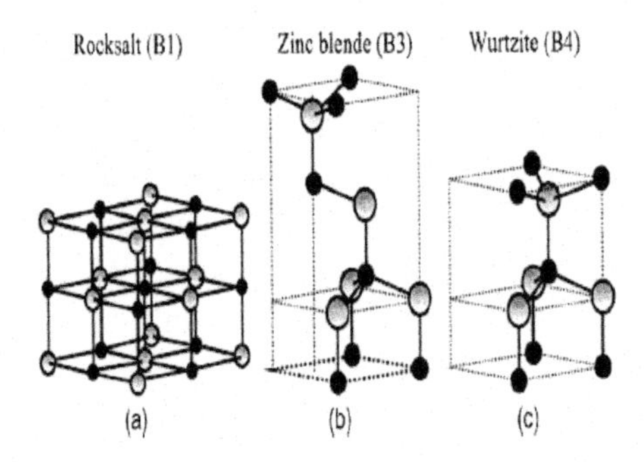

Fig. 1. Stick and ball representation of ZnO crystal structure; black and gray spheres denote Zn and O atoms respectively. [Reprinted with permission from U. Özgür et al., J. of Appl. Phys., 98, 041301 (2005).]

Bandgap engineering is available by adding Cadmium (Cd) or Magnesium (Mg) content to ZnO. The alloys $Zn_{(1-x)}Mg_xO$ (3.3 - 4 eV) and $Zn_{(1-y)}Cd_yO$ (2.9 - 3.37 eV) enable adjustments in optical properties with little variation of the semiconductor lattice parameter leading to heterojunctions with reduced strain.

ZnO nanomaterials have been studied for the benefits associated with quantum confinement including reduced dark noise, increased absorption efficiency and the potential for large area and lower cost devices. The nanostructures, however; often exhibit a strong, parasitic green photoluminescence caused by excess Zn^{2+} ions and oxygen deficiency[10,14,15,17]. Solutions to these problems include developing effective surface modification, annealing treatment and producing stable ZnO QDs that are unagglomerated, highly crystalline, and monodisperse[18-22]. The ability to create reproducible Ohmic or Schottky contacts is also associated with oxygen vacancies[23]. These vacancies often dominate which tend to pin the ZnO Fermi level to the defect level, hindering the expected barrier producing relationship between the contact metal and semiconductor.

The following sections will summarize and compare device results from bottom-up nanoparticle growth methods and an inexpensive alternative; the top down wet chemical etch method. The latter method addresses the issues associated with defect levels that often hinder performance.

3. Top-Down Wet-Chemistry ZnO Nanoparticle based UV Photodetectors

Wet chemical synthesis is both inexpensive and simple. Much of the wet-chemistry research can be classified as bottom-up synthesis. They are based on a range of precursors and synthesis conditions, such as temperature, time, and concentration of reactants, leading to different sizes and geometries of the resulting particles. However, several significant challenges remain, including difficulty in preparation of a stable dispersion, poor uniformity in the coating process, and poor conductivity of the ZnO layers.

The top-down approach was first reported by Sharma et al. in which bulk ZnO material was reduced to generate controllable and stable dispersions of ZnO nanoparticles through wet-chemistry[24]. This method enables an additional surface coating step with relative ease and uniformity. To investigate the effects of surface coating, the nanoparticles were coated with polyvinyl alcohol (PVA) (ZnO-U) and compared to uncoated ZnO nanoparticles (ZnO-A) produced by the same process. Figure 2 is a high resolution SEM image of PVA coated ZnO. The sizes range from 10 nm to 150 nm with an average value of 80 nm. The uncoated ZnO have a similar result.

Fig. 2. High resolution SEM result of PVA coated ZnO (ZnO-U) created by top-down wet-chemical synthesis with an average size of 80 nm.

3.1. *Parasitic green photoluminescence reduction and enhanced UV absorption*

Typical photoluminescence results of ZnO include UV emission centered near 380 nm and a dominant parasitic green emission peak. Gong et al. investigated the optical properties of ZnO spherical nanoparticles prepared by bottom-up synthesis by varying precursor mixtures, 1-D octadecene, triocylamine and trioctylphosphine, which modify the density of oxygen near the surfaces[25]. Borgohain and Mahammuni encapsulated ZnO nanoparticles with passivating agents such as polyvinyl prrolidone (PVP), propionic acid (PA) and tetra octyl ammonium bromide (TOAB) also grown through bottom-up methods[17]. The PVA coated top-down wet chemical etch produced nanoparticles derived from bulk ZnO[26].

Both uncoated and PVA coated ZnO nanoparticles were spin-cast on quartz and measured by photoluminescence. Figure 3 demonstrates the effectiveness of PVA coating to suppress parasitic green emission associated with defects in the material by acting as a surface passivation layer. The deep level surface traps are responsible for hindering band-to-band recombination. PVA has a strong interaction with excess Zn^{2+} ions during the top-down wet-chemistry process thereby reducing trapped carriers and enhancing the proportion of carriers contributing to band-to-band emission.

3.2. *Photoconductive and Metal-Semiconductor-Metal (MSM) device response*

Ohmic and Schottky contacts are ideally dependent upon the relative work functions of the contact metal and semiconductor. Metals for ohmic contacts with n-type ZnO include Aluminum (Al) and Indium (In) in addition to various metal combinations such as Titanium/Gold (Ti/Au), Aluminum/Platinum (Al/Pt)[27]. Candidates for Schottky contacts are Platinum (Pt), Gold (Au), Nickel (Ni), and Silver (Ag). Nevertheless, the surface defect states, residual surface contamination, and/or the interfacial gap between the metal and semiconductor often dominate creating some unpredictability in current-voltage (I-V) characteristics. For example, Jun et al. reported ohmic behaviour or linear I-V response with Au metal deposition on the n-type ZnO nanoparticles. Typically a non-linear response is expected but surface impurities and defects of ZnO are cited[8,28-33].

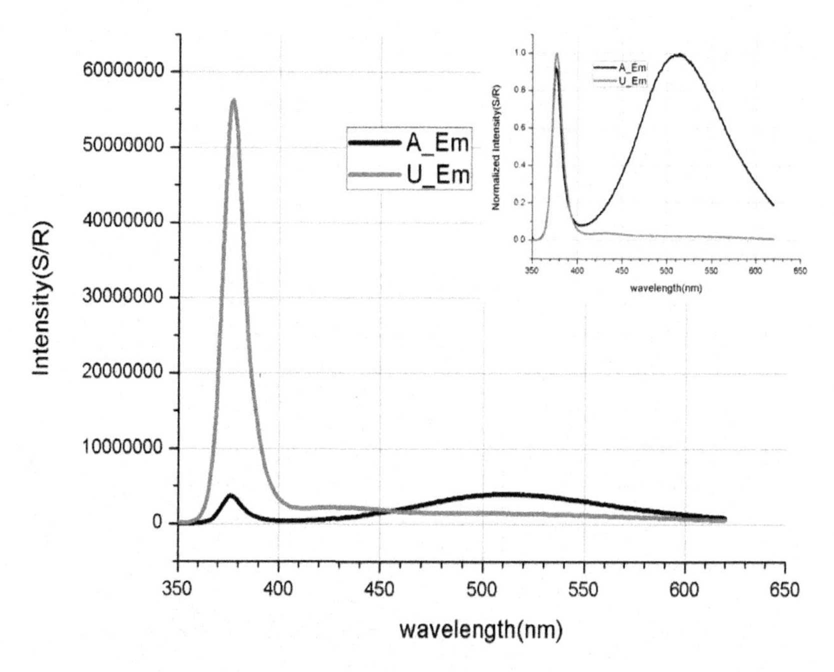

Fig. 3. Emission spectra of ZnO coated with PVA (U_Em) and uncoated (A_Em) dispersed in ethanol and then spin-cast on quartz. Strong UV emission peak at 377 nm from the band edge of emission. PVA coating nearly eliminates parasitic effect. The inset is the normalized spectra.

The effects of surface passivation also contribute to I-V response. The general mechanism of photocurrent for ZnO nanoparticles begins with adsorbed oxygen molecules on the surface as negatively charged ions, capturing free electrons [$O_2(g) + e^- \rightarrow O_2^-(ad)$] producing a depletion layer. Incident light with energy higher than the bandgap creates electron and hole pairs. The generated holes near the surface, neutralize chemisorbed oxygen [$h^+ + O_2^-(ad) \rightarrow O_2(g)$]. The depletion region narrows and photocurrent increases. Traps due to surface defects hinder the conducting mechanism. To investigate the photoconductivity of ZnO nanoparticles and verify the effect of surface passivation, photodetectors based on ZnO-A and ZnO-U were made respectively. A schematic diagram of the detectors is shown in the inset of Figure 4. The typical I-V plots of detectors with Al contacts above both ZnO-A and ZnO-U nanoparticle films in the dark and under 340 nm UV LED illumination with the same intensity (45.58mW/cm^2) are shown in Figure 4. The linear I-V plots demonstrate photoconductive detectors based on ZnO nanoparticles. The contacts were made by depositing two irregular 100 nm Al contacts via e-beam lithography and standard lift-off process. These measurements were performed at room temperature in air. The UV photo-generated current to dark current ratio (on/off ratio) in detector based on ZnO-A is about 9×10^3 when the bias is 20V, while for detector based on ZnO-U, the ratio is as high as 5×10^4, more than 5 times of that based on ZnO-A. These results are due to the higher defect concentration in ZnO-A compared to ZnO-U. Most of the photo-generated carriers are trapped by the defects during their transport to the terminals. However, the dark currents of both detectors are about 100 pA at 20V bias, which did not change with the PVA coating. This indicates PVA does not change the conductivity of ZnO nanoparticles, but acts as an effective surface passivation material to improve the photosensitivity of ZnO nanoparticles. In comparison to ZnO thin film photodetectors, the dark current of the above nanoparticle based detectors are 3 orders of magnitude less than the thin film ZnO photodetectors[34-36]. Jandow et al., Jiang et al. and Liu et al, demonstrated 1, 2, and 4 orders of magnitude between UV photogenerated current to dark current respectively[34-36].

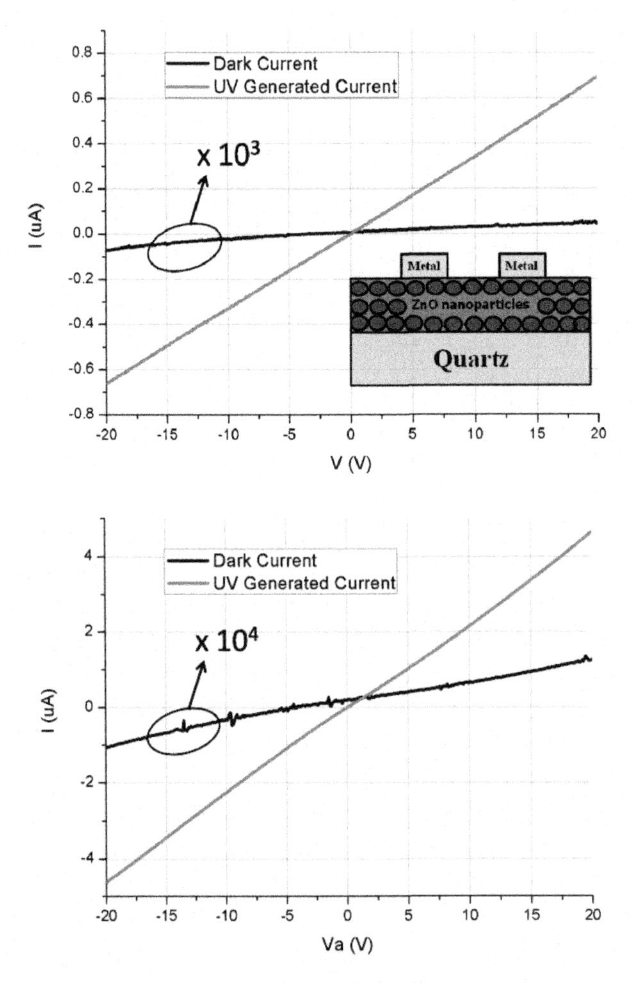

Fig. 4. Linear I-V response with Al contacts of (Top) Uncoated ZnO (ZnO-A) with UV/dark ratio of 9×10^3 at 20V bias (Bottom) PVA coated (ZnO-U) with UV/dark ratio of 5×10^4 at 20V bias.

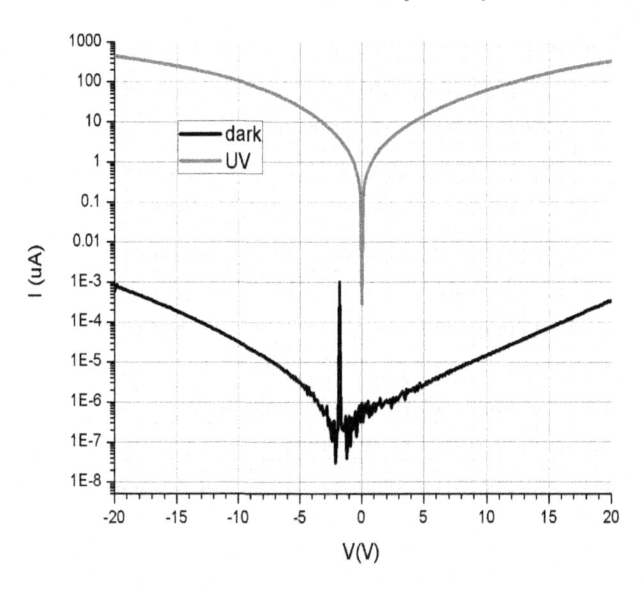

Fig. 5. PVA coated (ZnO-U) with Gold contacts. Non-linear I-V response with UV/dark ratio of 1×10^6 at 20V bias.

Figure 5 shows a non-linear I-V response of Au contacts deposited through optical lithography methods on PVA coated ZnO nanoparticles (ZnO-U) on quartz. Interdigitated finger contact arrays were designed with varying finger spacing, finger lengths, and finger widths. Each tested device demonstrated an MSM, non-linear response similar to Figure 5. When the detector with Au contacts was illuminated by a 340nm UV LED (45.58mW/cm^2), the UV generated current to dark current ratios were between 4 and 7 orders of magnitude.

In contrast to detectors with Al contacts deposited by e-beam through shadow mask, the UV generated current surpassed 100 uA at 20V bias for most MSM detectors which is an improvement of two orders of magnitude, and the dark current is about one order higher. Au contacts deposited by optical lithography rather than through a shadow mask via e-beam lithography may have reduced distance between contacts, thus decreasing the recombination possibility, which results in a higher UV generated current to dark current ratio.

The photoresponse of the ZnO nanostructures are on the order of seconds or tens of seconds in literature. Jun et al. reported rise and decay times of 48s and 0.9s respectively for 20-150 nm solution processed nanoparticles[8]. Soci et al. reported rise and decay times of 23s and 33s respectively for 150-300 nm diameter nanowires created by chemical vapor deposition[33]. Kind et al. found 1s rise and decay times for 20-300 nm diameter nanowires[31].

The time response results for PVA coated ZnO nanoparticles with irregular Al contacts are shown in Figure 6. The samples were illuminated by a 340 nm LED with the intensity of 4.58mW/cm^2. The results of 22s rise and 11s decay times are consistent with that of nanostructures with UV generated current to dark current ratios near the same order of magnitude. The rise and decay of the photocurrent is related to the adsorption of oxidizing molecules in ambient air and the time required for oxygen molecule diffusion. Similar measurements will be explored for the samples with Au MSM contact structures.

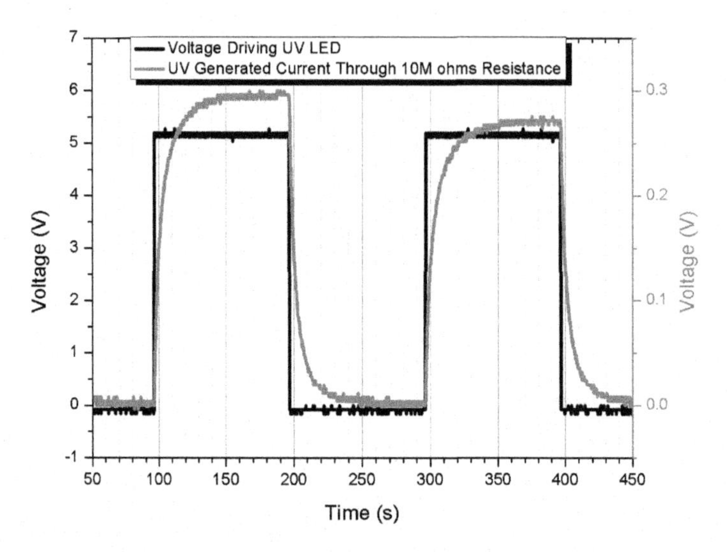

Fig. 6. Photocurrent time response measurement with 340 nm LED excitation.

3.3. *ZnO nanoparticles deposited on GaN-based substrates*

Wavelength selective photodetctors in the UV range were created with PVA coated ZnO nanoparticles deposited on GaN and AlGaN epitaxial substrates grown by Sensor Electronic Technology Inc (SET Inc.). The spectral response of these detectors is shown in Figure 7. When light penetrates the detector through the substrate or backside of the detector, a bandpass response related to the cut-off wavelength of substrate and ZnO nanoparticles is achieved. As shown in Figure 7, the absorption cut-off wavelengths of AlGaN and GaN are 300nm and 360nm, respectively through engineered epitaxial layer growth. When the wavelength of the light is shorter than the cut-off wavelengths, the light is absorbed by the substrate material and electron and hole pairs are generated. The photogenerated carriers do not have a conduction path to the contacts that are deposited on ZnO nanoparticles, thus current is not generated. Wavelengths longer than the substrate cutoff, penetrates the substrate and reaches the ZnO active layer which forms a MSM photodetector. If the wavelength is longer than the cut-off wavelength of ZnO, the light is not absorbed by ZnO and current is not generated. This is a bandpass response. The expected normal short pass response is also shown in Figure 7 when the light penetrates from the front side. In this case, the calculated responsivity of the ZnO nanoparticle-AlGaN substrate device is 0.8A/W at 375nm, and the spectral response is twice that of a commercial Si UV enhanced photodetector measured in the same system.

Wavelength selective detectors can be tuned by the material properties of both the substrate and the nanoparticles.

4. Summary

ZnO is an emerging material for blue and UV optoelectronics. Its properties are similar to GaN materials but differ in its low cost, reduced lattice strain from layered deposition of its alloys, larger excitonic binding energy, and availability of large-area native substrates. However, its prevalence is hindered by the lack of a reliable and reproducible p-type material. ZnO nanostructures introduce benefits associated with quantum confinement including reduced dark noise, increased absorption efficiency and the potential for large area and lower cost devices. High quality nanoparticles were created with a top-down wet-chemical process and coated with PVA. As a result, parasitic green photoluminescence, caused by deep level surface traps, was nearly eliminated. PVA provides surface passivation by reducing the number of excess Zn^{2+} ions. Photoconductive devices were created with both PVA coated and uncoated ZnO nanoparticles on quartz with Al contacts. A linear I-V response was achieved as expected. The PVA coated ZnO nanoparticles had a UV generated current to dark current ratio 5 times higher than the uncoated ZnO nanoparticles. The transient photoresponse from 340 nm LED excitation showed a rise and decay time of 22s and 11s respectively. These values are consistent with literature. Metal-semiconductor-metal devices were created by depositing Au interdigitated finger contacts through optical lithography. The UV generated current to dark ratio increased to seven orders of magnitude. Finally, wavelength selective devices were created by depositing PVA coated ZnO nanoparticles on epitazially grown AlGaN and GaN substrates. The absorption cutoff of AlGaN (300 nm) and GaN (360 nm) created a bandpass response when illuminated from through the substrate or backside of the detector. The responsivity

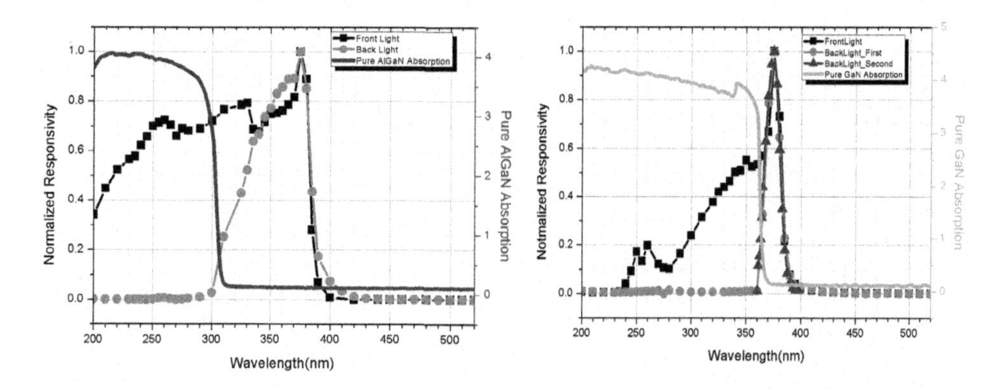

Fig. 7. Front and back illuminated UV detectors (left) ZnO nanoparticles on AlGaN substrate, shortpass (black line) and bandpass (red line: 300-380 nm) responsivity as it relates to the absorption of AlGaN (blue line) (right) ZnO nanoparticles on GaN substrate, shortpass (black line), and bandpass (red line: 360-380 nm), as it relates to the absorption of GaN (green line).

of the device with an AlGaN substrate was 0.8 A/W at 375 nm twice that of a commercial UV enhanced Si photodetector. These results indicate that PVA coating of ZnO nanoparticles during top-down wet-chemistry synthesis could be applied in creating low cost, sensitive, visible blind, and wavelength selective UV photodetectors.

Acknowledgments

The authors gratefully acknowledge support from NSF Industry/University Cooperative Research Center Connection One. We'd like to thank Professor Partha Dutta in Electrical, Computer and Systems Engineering Department Rensselaer Polytechnic Institute and Sensor Electronic Technology Inc. Finally, we would like to thank National Security Technologies for their support.

References

1. V. A. Karpina, V. I. Lazorenko, C. V. Lashkarev, V. D. Dobrowolski, L. I. Kopylova, V. A. Baturin, S. A. Pustovoytov, A. J. Karpenko, S. A. Eremin, P. M. Lytvyn, V. P. Ovsyannikov, and E. A. Mazurenko, Zinc oxide - analogue of GaN with new perspective possibilities, *Crystal Research and Technology*, **39**, 980–992, (2004).
2. Z. Bi, J. Zhang, X. Bian, D. Wang, X. Zhang, W. Zhang, and X. Hou, A high-performance ultraviolet photoconductive detector based on a ZnO film grown by RF sputtering. *Journal of Electronic Materials*, **37**, 760–763, (2008).
3. G. Cheng, Z. Li, S. Wang, G. Gong, K. Cheng, X. Jiang, S. Zhou, Z. Du, T. Cui, and G. Zou, The unsaturated photocurrent controlled by two-dimensional barrier geometry of a single ZnO nanowire schottky photodiode, *Applied Physics Letters*, **93**, 123103.1–123103.3, (2008).
4. Y. Jin, J. Wang, B. Sun, J. Blakesley, and N. Greenham, Solution-processed ultraviolet photodetectors based on colloidal ZnO nanoparticles, *Nano Letters*, **8**, 1649–1653, (2008).
5. Y. Lin, C. Chen, W. Yen, W. Su, C. Ku, and J. Wu, Near-ultraviolet photodetector based on hybrid polymer/zinc oxide nanorods by low-temperature solution processes, *Applied Physics Letters*, **92**, 233301.1–23301.3, (2008).
6. S. J. Young, L. W. Ji, S. J. Chang, S. H. Liang, K. T. Lam, T. H. Fang, K. J. Chen, X. L. Du, and Q. K. Xu, ZnO-based MIS photodetectors, *Sensors and Actuators A: Physical*, **141**, 225–229, (2008).
7. D. C. Look, Zinc Oxide Bulk, Thin Films and Nanostructures, Elsevier, 21–42, (2006).
8. J. Jun, H. Seong, K. Cho, B. Moon, and S. Kim, Ultraviolet photodectors based on ZnO nanoparticles, *Ceramics International*, **35**, 2797–2801, (2009).
9. S. Lee, Y. Jeong, S. Jeong, J. Lee, M. Jeon, and J. Moon, Solution-processed ZnO nanoparticle-based semiconductor oxide thin-film transistors, *Superlattice Microstructures*, **44**, 761–769, (2008).
10. S. Monticone, R. Tufeu, and A. Kanaev, Complex nature of the UV and visible fluorescence of colloidal ZnO nanoparticles, *Journal of Physical Chemistry B.*, **102**, 2854–2862, (1998).
11. E. Neshataeva, T. Kummel, G. Bacher, and A. Ebbers, All-inorganic light emitting device based on ZnO nanoparticles, *Applied Physics Letters*, **94**, 091115.1–091115.3, (2009).
12. Y. Qin, X. Wang, and Z. Wang, Microfibre-nanowire hybrid structure for energy scavenging, *Nature*, **45**, 809–813, (2008).
13. M. Wang, Y. Lian, and X. Wang, PPV/PVA/ZnO nanocomposite prepared by complex precursor method and its photovoltaic application, *Current Applied Physics*, **9**, 189–194, (2009).

14. L. Wu, Y. Wu, X. Pan, and F. Kong, Synthesis of ZnO nanorod and the annealing effect on its photoluminescence property, *Optical Materials*, **28**, 418–422, (2006).
15. Y. Wu, A. Tok, F. Boey, X. Zeng, and X. Zhang, Surface modification of ZnO nanocrystals. *Applied Surface Science*, **253**, 5473–5478, (2007).
16. X. Zhua, I. Yurib, X. Gana, I. Suzukib, and G. Lia, Electrochemical study of the effect of nano-zinc oxide of microperoxidase and its application to more sensitive hydrogen peroxide biosensor preparation, *Biosensors and Bioelectronics*, **22**, 1600–1604, (2007).
17. K. Borgohain and S. Mahamuni, Luminescence behavior of chemically grown ZnO quantum dots, *Semiconductor Science and Technology*, **13**, 1154–1157, (1998).
18. L. Guo, S. Yang, C. Yang, P. Yu, J. Wang, W. Ge, and G. Wong, Highly monodisperse polymer-capped ZnO nanoparticles: Preparation and optical properties, *Applied Physics Letters*, **76**, 2901–2903, (2000).
19. K. Kim, N. Koguchi, Y. Ok, T. Seong, and S. Park, Fabrication of ZnO quantum dots embedded in an amorphous oxide layer, *Applied Physics Letters*, **84**, 3810–3812, (2004).
20. S. Mridha, M. Nandi, A. Bhaumik, and D. Basak, A novel and simple approach to enhance ultraviolet photosensitivity: activate carbon-assisted growth of ZnO nanoparticles, *Nanotechnology*, **19**, 275705, (2008).
21. H. Xiong, Z. Wang, and Y. Xia, Polymerization initiated by inherent free radicals on nanoparticle surfaces: a simple method of obtaining ultrastable (ZnO) polymer core-shell nanoparticles with strong blue fluorescence, *Advanced Materials*, **18**, 748–751, (2006).
22. S. Yang and C. Park, Facile preparation of monodisperse ZnO quantum dots with high quality photoluminescence characteristics, *Nanotechnology*, **19**, 035609.1–035609.4, (2008).
23. M. W. Allen and S. M. Durbin, Influence of oxygen vacancies on schottky contacts to ZnO. *Applied Physics Letters*, **92**, 122110–122113, (2008).
24. S. Sharma, A. Tran, O. Nalamasu, and P. Dutta, Spin coated ZnO thin films using ZnO nano-colloid, *Journal of Electronic Materials*, **35**, 1237–1240, (2006).
25. Y. Gong, T. Andelman, G. F. Neumark, S. O'Brien, and I. L. Kuskovsky, Origin of defect-related green emission from ZnO nanoparticles: effect of surface modification, *Nanoscale Research Letters*, **2**, 297–302, (2007).
26. L. Qin, C. Shing, and S. Sawyer, Ultraviolet photodetection based on ZnO colloidal nanoparticles made by top-down wet-chemistry synthesis process, accepted Symposium on Photonics and Optoelectronics, (2010).
27. J. H. Lim and S. J. Park, *Zinc Oxide Bulk, Thin Films and Nanostructures*, Elsevier, 267–293, (2006).
28. S. E. Ahn, J. S. Lee, H. Kim, S. Kim, B. H. Kang, and K. H. Kim, Photoresponse of sol-gel-synthesized ZnO nanorods, *Applied Physics Letters*, **84**, 5022–5024, (2004).
29. B. J. Coppa, C. C. Fulton, S. M. Kiesel, R. F. Davis, C. Pandarinath, J. E. Burnette, R. J. Nemanich, and D. J. Smith, Structural, microstructure, and electrical properties of gold films and schottky contacts on remote place-cleaned, n-type ZnO 0 0 0 1 surfaces, *Journal of Applied Physics*, **97**, 103517.1–103517.13, (2005).
30. S. S. Hullavarad, N. V. Hullavarad, M. Mooers, and P. C. Karulkar, Fabrication of nanostructured ZnO UV sensor, In *Materials Research Society Symposium Proceedings*, **951**, pages 216–220, (2007).
31. H. Kind, H. Yan, B. Messer, M. Law, and P. Yang, Nanowire ultraviolet photodetectors and optical switches, *Advance Materials*, **14**, 158–160, (2002).
32. S. Liang, H. Sheng, Y. Liu, Z. Huo, Y. Lu, and H. Shen, ZnO schottky ultraviolet photodetectors, *Journal of Crystal Growth*, **225**, 110–113, (2001).
33. C. Soci, A. Zhang, B. Xiang, S. A. Dayeh, D. P. R. Aplin, J. Park, X. Y. Bao, Y. H. Lo, and D. Wang, ZnO nanowire UV photodetectors with high internal gain, *Nano Letters*, **7**, 1003–1009, (2007).

34. N. N. Jandow, K. A. Ibrahim, H. A. Hassan, S. M. Thahab, and O. S. Hamad, The electrical properties of ZnO MSM photodetector with Pt contact electrodes on PPC plastic, *Journal of Electron Devices*, **7**, 225–229, (2010).
35. D. Jiang, J. Zhang, Y. Lu, K. Liu, D. Zhao, Z. Zhang, D. Shen, and X. Fan, Ultraviolet schottky detector based on epitaxial ZnO thin film, *Solid-State Electronics*, **52**, 679–682, (2008).
36. Y. Liu, G. R. Gorla, S. Liang, N. Emanetoglu, Y. Lu, H. Shen, and M. Wraback, Ultraviolet detectors based on epitaxial ZnO films grown by MOCVD, *Journal of Electronic Materials*, **29**, 69–74, (2000).

CARBON-BASED NANOELECTROMECHANICAL DEVICES

STEFAN BENGTSSON, PETER ENOKSSON, FARZAN A. GHAVANINI, KLAS ENGSTRÖM, PER LUNDGREN

Department of Microtechnology and Nanoscience, Chalmers University of Technology, 412 96 Göteborg, Sweden
stefan.bengtsson@chalmers.se

ELEANOR E. B. CAMPBELL, JOHAN EK-WEIS

School of Chemistry, University of Edinburgh, West Mains Road, Edinburgh, Scotland EH9 3JJ
eleanor.campbell@ed.ac.uk

NIKLAS OLOFSSON, ANDERS ERIKSSON

Department of Physics, University of Gothenburg, 412 96 Göteborg, Sweden
niklas.olofsson@physics.gu.se

Carbon-based nanoelectromechanical devices are approaching applications in electronics. Switches based on individual carbon nanotubes deliver record low off-state leakage currents. Arrays of vertically aligned carbon nanotubes or nanofibers can be fabricated to constitute varactors. Very porous, low density arrays of quasi-vertically aligned arrays of carbon nanotubes behave mechanically as a single unit with very unusual material properties.

Keywords: carbon nanotubes; carbon nanofibers; nanoelectromechanical systems; switch; varactor.

1. Introduction

The theoretical model predictions and experimental observations of actual response to electrical and mechanical stimuli of the carbon nanotube material family show that it is worthwhile to design and fabricate electronic devices using such materials. Advances for employing carbon nanotubes (CNTs) as conductors of current and/or heat in future integrated electronic systems have been extensively reported on and progress in that area has been rapid[1-3]. CNTs are also investigated in the role of acting as the active element in transistors[4,5]. With a broad perspective on electronics, carbon nanostructures have been demonstrated as an attractive choice as electrode material in sensor applications[6]. A different class of devices exploits the interplay between electrical and mechanical effects as the very key to the device functionality in a nanoelectromechanical system (NEMS). Although it is possible to conceive of many different devices and applications of NEMS made up of CNTs or similar materials, we focus our description on switches and varactors as model examples of such devices in this overview of recent advances towards the realization of carbon-based NEMS.

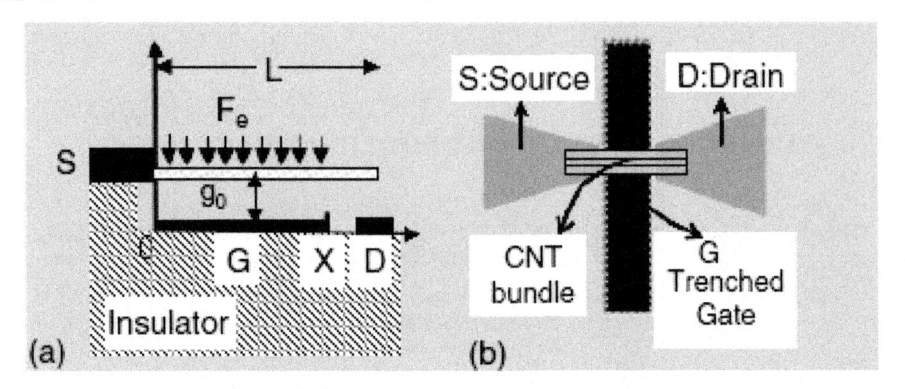

Fig. 1. Schematic pictures of the (a) singly clamped and (b) doubly clamped configurations for a beam switch. From Yousif[10] reprinted with permission from IOP Publishing.

2. Modeling

By importing theoretically predicted[7] or measured values for the Young's modulus of CNTs[8] or carbon nanofibers (CNFs)[9] and by employing continuum mechanics in conjunction with analytical expressions or boundary element methodology we can couple the mechanical actuation to voltage induced electrical forces. In this way it is possible to make predictions of the behavior of electronic devices like switches and varactors. In the case of a switch, the basic functionality is to change conductivity from infinite to zero as fast as possible and with minimum cost of energy. One attractive feature of using CNT electromechanics to realize a switching device is that the small dimensions and high stiffness will give a high resonance frequency. This enables fast switching. Furthermore,

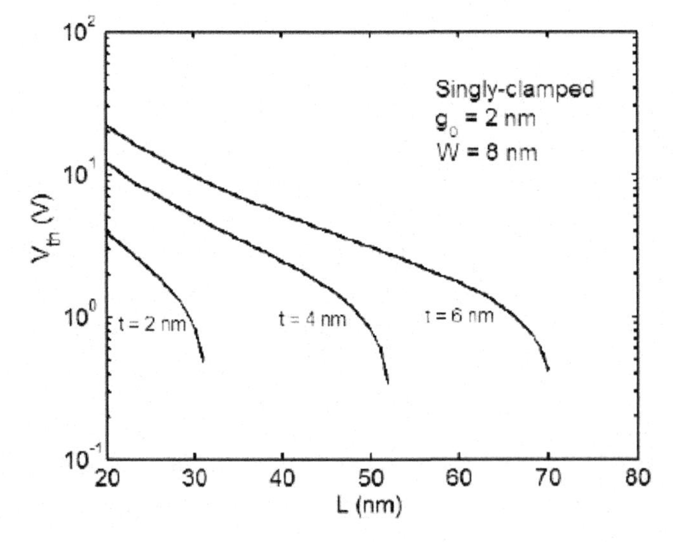

Fig. 2. The dependence of the threshold voltage on the length of the switching beam. From Yousif[10] reprinted with permission from IOP Publishing.

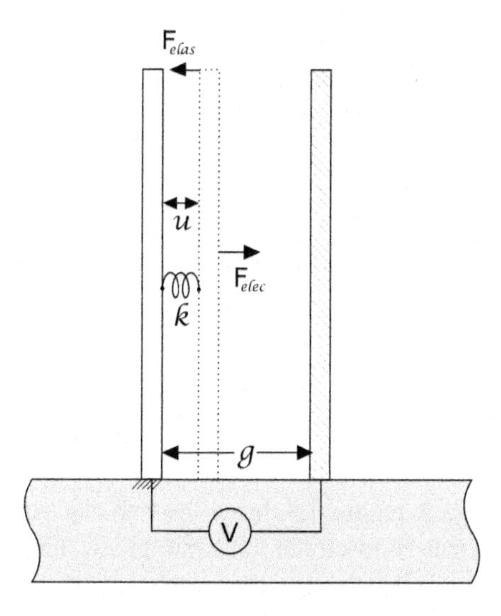

Fig. 3. One-dimensional lumped model of the CNF varactor.

the disjunct off-state of such a device will yield a leakage current that will give negligible contribution to the total power consumption of the device[10].

In the case of varactors the functionality is more complex than just achieving a transition from an on-state to an off-state. The geometry considered in this paper employs a pair of nanoelectromechanical electrodes to obtain the voltage dependent capacitance characteristic of varactors. These non-linear circuit elements find use e. g. in voltage controlled oscillators, and some of the critical features of the varactors are their capacitance per unit chip area and their swing in capacitance for the applicable voltage range. The modeling methods employed to describe these devices comprise boundary element calculations to resolve the geometrical effects on the nanostructures as well as simplistic analytical descriptions which can capture the qualitative behavior and give first order estimates of critical parameters.

2.1. *Switches*

In order to evaluate the optimal performance of digital switching devices fabricated using carbon nanotubes, a continuum mechanics approximation has been employed to describe mechanically switching beams of either multi-wall carbon nanotubes (MWCNT) or dense bundles of single-wall carbon nanotubes (SWCNT)[10]. The thickness of the beam, t, is a crucial design parameter among the geometrical dimensions where the beam length, L, and the nominal distance between beam and the actuating electrode, g_0, also play an important role. Figure 1 shows a schematic illustration of the geometrical configurations considered in the paper by Yousif et al.[10] and in Figure 2 the calculated threshold voltage to turn the switch on is displayed as a function of the beam length with the thickness as parameter for a given beam-contact gap of 2 nm and a beam width, W, of 8 nm. In order to optimize the geometry it is important to be able to control the beam thickness, i. e. the nanotube diameter or in the case of bundles, the number of nanotubes. Still, exploiting the freedom of the design space and assuming a high Young's modulus of 1 TPa, the carbon nanotube nanoelectromechanical switching will be more than one order of magnitude slower than a DRAM element at the same critical length (gate length and beam length). On the up side, the NEMS switch will consume orders of magnitude less power, mainly due to negligible off-state leakage current (1×10^{-5} $\mu A\mu m^{-1}$ for CMOS DRAM and 1×10^{-9} $\mu A\mu m^{-1}$ for the CNT switch). With a low actuation voltage design, the energy cost for a switching event can be in the aJ range[10].

2.2. *Varactors*

In the case of varactors one focus issue of the modeling we have employed is to correlate the effective Young's modulus of the NEMS with the experimentally measurable pull-in voltage – the minimum voltage required to bring the two electrodes of the capacitor into contact, forming an electrical short-circuit. In a simplistic lumped capacitance model, depicted in Figure 3, it is possible to arrive at an analytical relationship for the pull-in voltage as a function of the geometrical dimensions of the actuated electrode and of its effective spring constant, k[11]. Such a model is not powerful enough to faithfully resolve the interplay between pull-in voltage, geometry and stiffness of an individual NEMS element to the extent that we can understand e. g. the impact of growth conditions on its experimentally observed mechanical behavior. Using the boundary element method (BEM) in simulations[12] it is possible to obtain the deflection at a given geometry and biasing condition in an iterative fashion; the charge distribution is calculated first, then the resulting electric force, thereafter the deflection, after which the charge distribution (at the given voltage) is recalculated to start the next round of iteration. The outcome of such a simulation is displayed in Figure 4. In a design space delimited by estimations of

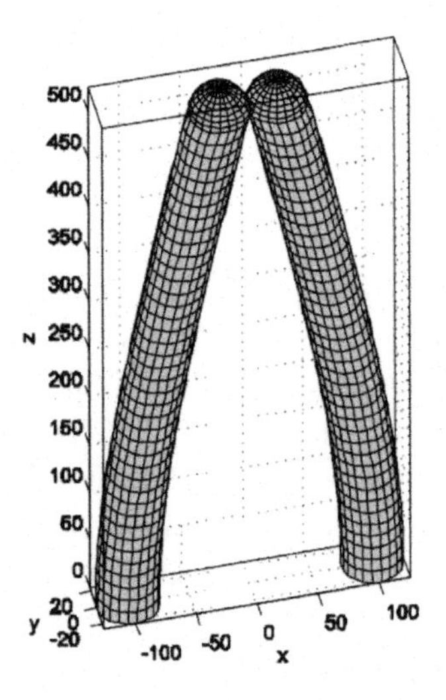

Fig. 4. This image shows the result of applying the BEM to calculate the nanofiber deflection for a given applied voltage.

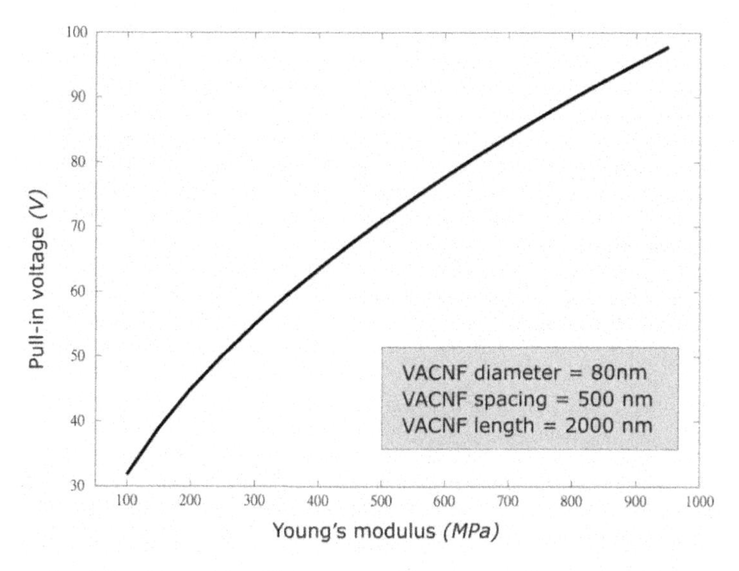

Fig. 5. Resulting relationship between pull-in voltage and Young's modulus.

experimentally relevant geometries, voltages and Young's moduli, the BEM simulations are run to generate a database of correlations between pull-in voltage and Young's modulus; a measurement of the pull-in voltage can then be directly correlated to a value of the Young's modulus for a given experimental condition, by referring to a generated relationship such as the one shown in Figure 5.

3. Device Fabrication and Characterization

The technique of chemical vapor deposition (CVD) can produce CNTs and CNFs with a desired functional pattern. Any electronic function designed on a chip carries a cost in terms of its footprint, which makes it immediately favorable to design NEMS devices that extend upward, perpendicular to the chip surface. Depending on the processing details it is possible to obtain either individual free-standing vertically aligned CNFs (Figure 6) or

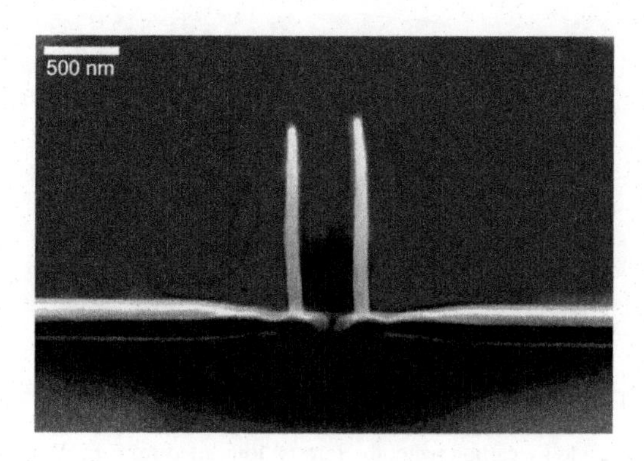

Fig. 6. An SEM picture of a pair of individually contacted CNFs separated by a distance below 500 nm.

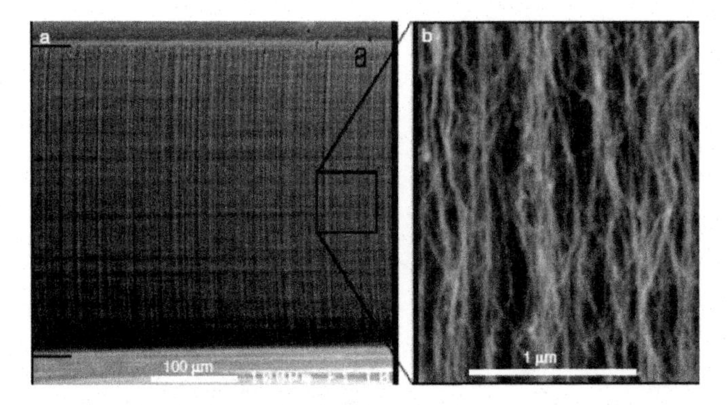

Fig. 7. The forest of CNTs with a close up view to show the wiggly and porous structure. From Olofsson[16] reprinted with permission from IOP Publishing.

forests of erect CNTs (Figure 7). In the case of CNFs the electric field of a plasma environment controls the growth directionality[13], whereas it is the interaction between CNTs grown at a high enough density and rate which induces the vertical alignment in the forest configuration[14]. The CNFs have a complex internal structure with less order than a nanotube[9], but one advantage is that individual vertically aligned nanostructures can be grown. It is not possible to achieve this with individual carbon nanotubes.

3.1. *Individual Vertically Aligned Carbon Nanofibers*

Electron beam patterned nanoscale Ni dots have been employed to catalyze the growth of nanofibers with a diameter of 100 nm and lengths exceeding 1 µm. Plasma CVD with a dc-plasma at a current of 20 mA was used to grow the fibers at 700 °C. The CNF pair of Figure 6 is an example of best achievement. The insulation between the individual fibers appears to be very good, with sub 0.1 nA current leakage up to 100 V. There is still need for further process optimization before we can fabricate single carbon nanofibers deterministically as a NEMS building block, but this development is on the level of optimization for specific processing tools, with the main hurdle being the choice of least

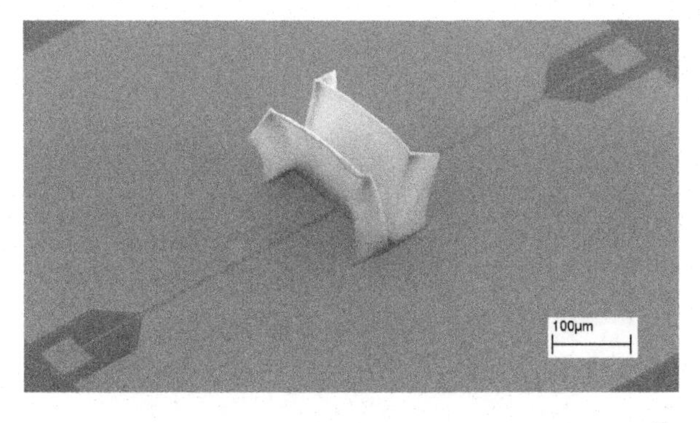

Fig. 8. Varactor electrodes made by carbon nanotube forests. Reprinted from Ek-Weis[17] with permission from Professional Engineering Publishing.

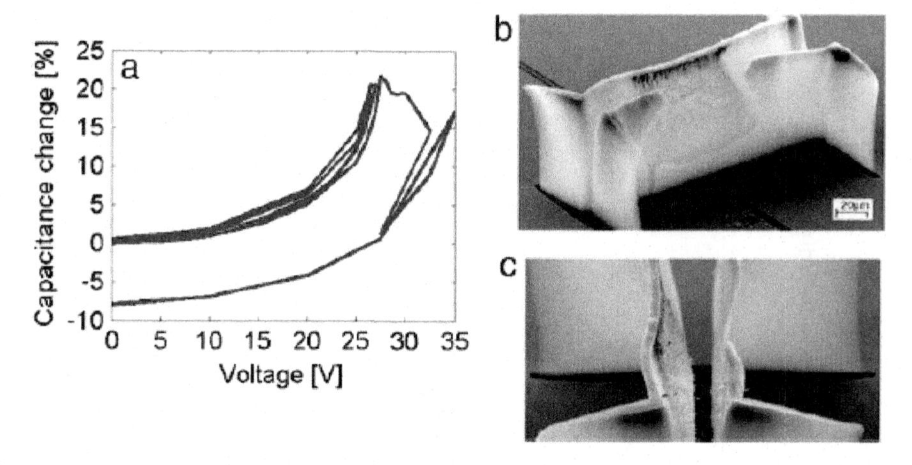

Fig. 9. a) Capacitance change as a function of voltage; b) and c) SEM images of the structure after exceeding pull-in. From Olofsson[16] reprinted with permission from IOP Publishing.

interfering work-around to handle the discharging instabilities in the dc-plasma. Individual electrically addressable vertically aligned CNFs without mechanical degrees of freedom have been demonstrated[15].

3.2. *Carbon Nanotube Forest*

Using thermal CVD at 700 °C with Fe catalyst, forests of vertically aligned 135 μm high multiwalled carbon nanotubes have been grown with a length of 200 μm, a width of 4 μm and a lateral separation of 10 μm[16]. The areal density of nanotubes in the forest was estimated at 10^{10} nanotubes cm^{-2}, and matching simulations to the measured actuation of these varactor electrodes yielded an effective Young's modulus on the order of a few MPa, i. e. far below the TPa often attributed to individual nanotubes, thus allowing actuation to be achieved for relatively low applied voltages[16]. The very low effective Young's modulus can be attributed to the highly porous and "wiggly" nature of the material (Figure 7). A varactor device designed using this material is displayed in Figure 8, showing the buttresses that need to be added to achieve sufficient mechanical rigidity. In Figure 9 the degree to which capacitance tuning can be achieved is shown along with the consequence of exceeding the pull-in voltage, as illustrated by two SEM images after the catastrophic event, which also alters the capacitance-voltage characteristic. The capacitance was determined by matching S parameter simulations to measurements, using an equivalent circuit which gave very good agreement with experimental observations[17].

3.3. *CMOS Compatibility*

It is a very attractive goal to be able to harness the advantages of carbon-based NEMS on a conventional CMOS electronics platform. Integration is a crucial issue when aiming for competitive system level performance for devices that incorporate and utilize NEMS.

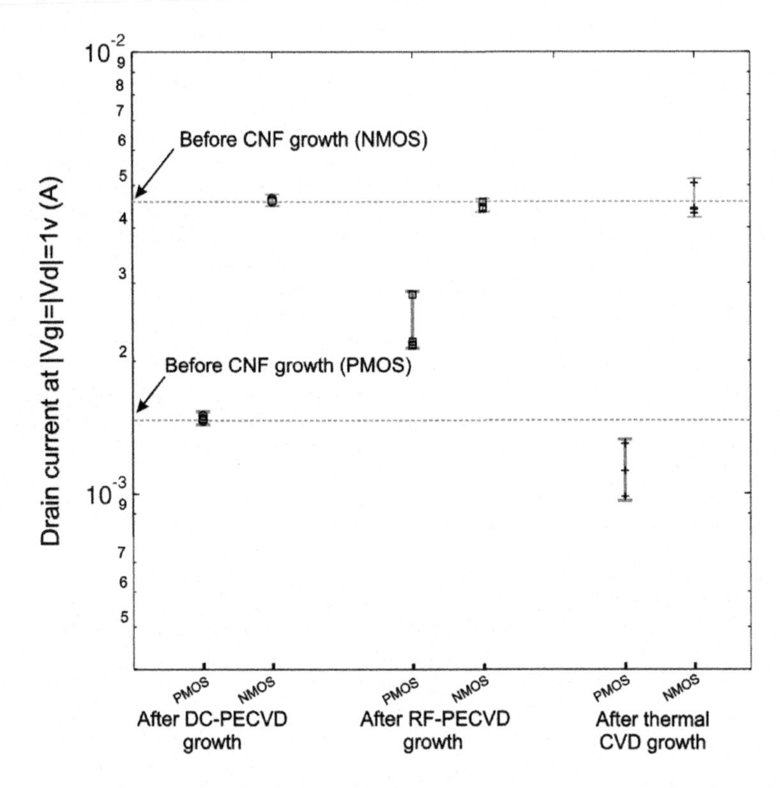

Fig. 10. Deterioration of the on-state drain current of PMOS and NMOS transistors after three disparate growth processes. From Ghavanini[18]; reprinted with permission from ACS Publications.

However, finding a way to match the processing requirements to obtain good carbon nanostructures with the restrictions for processes and materials in CMOS production is not trivial. The outcome of exposing transistors fabricated in 130 nm bulk CMOS technology to carbon nanofiber growth conditions, i. e. elevated temperatures and a plasma environment, shows that transistors can survive such a treatment and even perform without immediate detrimental consequences. Comparing rf-plasma processing at 560 °C, thermal CVD at 610 °C and exposure to a dc-plasma at 500 °C, the last of these three nanofiber growth methods gives the least impact on transistor performance[18], and in the case of the on-state drain current there is no discernible effect, as is shown in Figure 10.

4. Conclusion

In theory, carbon-based nanoelectromechanical switches can provide a low static power technology with potential for high frequency operation. Components constituted by carbon nanotubes will display high mechanical stability and will also be highly stable with regards to temperature. However, the dependencies of device properties like these on the specific growth conditions need further investigations. Large scale reproducible and reliable manufacturing of integrated NEMS elements remains extremely challenging, where growth on top of CMOS as a back-end process further increases the complexity and limits the degrees of freedom by having to resolve the conflict between CMOS compatibility requirements and carbon nanostructure quality; this integration scenario is still however a future possibility albeit not obviously achievable.

CNT-based NEMS switches demonstrate very low off-state leakage currents. Varactors can be realized with individual vertically aligned CNFs or with walls consisting of quasi-vertically aligned arrays of sparse CNTs. The latter are very porous, but behave mechanically as a cohesive unit with exceptional material properties.

Acknowledgments

The authors would like to acknowledge the external providers of financial support for this work, namely the sixth framework for research funding of the European Commission (contract 003673 CANEL and contract 028158 NANORF) and the Swedish Foundation for Strategic Research (SSF).

References

1. A. P. Graham, G. S. Duesberg, R. Seidel, M. Liebau, E. Unger, F. Kreupl, and W. Honlein, Diamond and Related Materials **13**, 1296-1300 (2004).
2. J. Robertson, Materials Today **10**, 36-43 (2007).
3. P. Avouris and C. Jia, Materials Today **9**, 46-54 (2006).
4. P. Chan, Y. Chai, Z. Min, and F. Yunyi, in *The application of carbon nanotubes in CMOS integrated circuits*, Piscataway, NJ, USA, 2008.
5. M. Haselman and S. Hauck, Proceedings of the IEEE **98**, 11-38 (2010).
6. A. V. Melechko, R. Desikan, T. E. McKnight, K. L. Klein, and P. D. Rack, Journal of Physics D: Applied Physics **42**, 193001 (28 pp.) (2009).
7. S. Reich, C. Thomsen, and J. Maultzsch, *Carbon Nanotubes: Basic Concepts and Physical Properties*, 1 ed. (Wiley-VCH, Berlin, 2004).
8. C. W. S. To, Finite Elements in Analysis and Design **42**, 404-413 (2006).
9. A. V. Melechko, V. I. Merkulov, T. E. McKnight, M. A. Guillorn, K. L. Klein, D. H. Lowndes, and M. L. Simpson, Journal of Applied Physics **97**, 39 (2005).
10. M. Y. A. Yousif, P. Lundgren, F. Ghavanini, P. Enoksson, and S. Bengtsson, Nanotechnology **19**, 285204 (7 pp.) (2008).
11. M. Dequesnes, S. V. Rotkin, and N. R. Aluru, Nanotechnology **13**, 120-131 (2002).
12. F. París and J. Cañas, *Boundary element method : fundamentals and applications*, Oxford Univ. Press (1997).
13. V. I. Merkulov, A. V. Melechko, M. A. Guillorn, M. L. Simpson, D. H. Lowndes, J. H. Whealton, and R. J. Raridon, Applied Physics Letters **80**, 4816-4818 (2002).
14. G. Eres, A. A. Kinkhabwala, H. Cui, D. B. Geohegan, A. A. Puretzky, and D. H. Lowndes, *Molecular beamcontrolled nucleation and growth of vertically aligned single-wall carbon nanotube arrays*, The Journal of Physical Chemistry B **109**, 16684-16694 (2005).
15. M. A. Guillorn, T. E. McKnight, A. Melechko, V. I. Merkulov, P. F. Britt, D. W. Austin, D. H. Lowndes, and M. L. Simpson, Journal of Applied Physics **91**, 3824-8 (2002).
16. N. Olofsson, J. Ek-Weis, A. Eriksson, T. Idda, and E. E. B. Campbell, Nanotechnology **20** (2009).

17. J. Ek-Weis, A. Eriksson, T. Idda, N. Olofsson, and E. E. B. Campbell, *Radio-frequency characterization of varactors based on carbon nanotube arrays*, Journal Proceedings of the Institution of Mechanical Engineers, Part N: Journal of Nanoengineering and Nanosystems **222**, 111-115 (2009).
18. F. A. Ghavanini, H. Le Poche, J. Berg, A. M. Saleem, M. S. Kabir, P. Lundgren, and P. Enoksson, *Compatibility Assessment of CVD Growth of Carbon Nanofibers on Bulk CMOS Devices*, Nano Letters **8**, 2437-2441 (2008).

CHARGE PUDDLES AND EDGE EFFECT IN A GRAPHENE DEVICE AS STUDIED BY A SCANNING GATE MICROSCOPE

J. CHAE, H. J. YANG, H. BAEK, J. HA, Y. KUK

Department of Physics and Astronomy, Seoul National, University Gwanak-gu,
Seoul, 151-747, Korea
ykuk@phya.snu.ac.kr

S. Y. JUNG, Y. J. SONG, N. B. ZHITENEV, J. A. STROSCIO

Center for Nanoscale Science and Technology, National Institute of Standards and Technology,
100 Bureau Drive, Gaithersburg, MD 20988, USA

S. J. WOO, Y.-W. SON

School of Computational Korea Institute of Advanced Studies,
Dongdaemoon-gu, Seoul, 130-722, Korea

Despite the recent progress in understanding the geometric structures of defects and edges in a graphene device (GD), how such defects and edges affect the transport properties of the device have not been clearly defined. In this study, the surface geometric structure of a GD was observed with an atomic force microscope (AFM) and the spatial variation of the transport current by the gating tip was measured with scanning gate microscopy (SGM). It was found that geometric corrugations, defects and edges directly influence the transport current. This observation is linked directly with a proposed scattering model based on macroscopic transport measurements.

Keywords: Graphene Device; Scanning Gate Microscopy; Carrier Uniformity.

1. Introduction

Graphene has been widely studied due to scientific interests and because of its possible application to high-speed devices[1-4]. Ever since the successful separation of graphene layers, many unique physical properties of graphene have been reported, including its linear dispersion relationship, relativistic fermionic behavior in the conduction and the valence bands, two-dimensional electron gas (2DEG) behavior, and back-scatteringless tunneling[5-9]. Many related theoretical predictions have been confirmed by macroscopic transport measurements[2,3,10]. Despite the successful explanation of the physical properties, the correlation between the geometric structures of defects such as those caused by corrugations, defects and edges and the carrier scattering at these defects is not fully understood. For example, the mobility of graphene measured in a suspended GD or epitaxially grown graphene is reportedly as high as 200,000 cm/V•s[11-16]. However, a GD on a SiO_2 substrate has been measured in the range of 5,000~10,000 cm/V•s[17-19]. The difference is modeled by the carrier scattering by various defects on a GD on SiO_2 substrate but not on a suspended GD. Thus far, structural studies have been conducted using microscopic tools such as electron microscopy and scanning tunneling microscopy[16,20-25], while the transport properties have been measured macroscopically in a two- or four-terminal device with a back gate[2].

We report local carrier transport measurement results using SGM. SGM is a unique microscopic tool with which the local geometric and electronic structures and the transport property of an electronic device can be measured simultaneously[26-34]. An SGM uses a conducting tip to apply an electric field locally and measures the transport current through two or four contacts. It can use the same tip to measure the geometric structure in Atomic Force Microscopy (AFM) mode. The results of this experiment showed that the local geometrical defects indeed work as scattering centers in the transport measurement.

2. Transport Property of GD

Even before any transport measurement of a GD were reported, Ando predicted the characteristics of the transport properties[35]; perfect graphene only with short-range scatters would show no conductivity dependence on the carrier density because the scattering rate would be divergent as the carrier density goes to zero near the Dirac point. However, the long-range scatters would be dominant in the low-density limit such that the device becomes insulating at the Dirac point. Earlier transport results[2,3] reported that the conductivity is linearly dependent on the induced charge density by the back gate voltage, even in the range of $\pm 1\times 10^{13}$ cm^{-2}. This result suggests that the transport property of a GD is strongly dependent on the scattering mechanism by long-range scatters. With the progress of GD fabrication, a GD was processed as a suspended device in order to remove the substrate effects[14,15]. It was predicted that the carrier transport would be affected by scattering with charged impurities, short-range scatterers, mid-gap states, various phonon modes, surface corrugations, and defects in a GD[17-19,36-40]. In the case of the charged impurity potential at a high carrier density limit, the conductance is given by

$$\sigma = C\frac{e^2}{h}\left|\frac{n}{n_{imp}}\right|, \tag{1}$$

which is linearly dependent on the induced charge density[17,38]. The dimensionless constant C is ~10 to ~20. A recent experiment reported that the measured mobility is linearly dependent on the doping level of potassium[38,41].

Experimentally, the existence of charge puddle in graphene on the top of the SiO_2 substrate was resolved using a scanning single-electron transistor microscope[42]. That experimental result was considered in the theoretical calculation of transport properties of GD near Dirac point[43]. Recently, another investigation resolved the charge impurity scattering centers using a spatial map in scanning tunneling spectroscopy (STS). In that study, the Dirac point was locally mapped from STS data and the scattering pattern was analyzed to determine the scattering centers. Good agreement was noted between the local Dirac point and a theoretical model of carrier scattering[44].

Another important scattering mechanism in a GD is scattering with a phonon. Raman spectroscopy results in the presence of an electric field showed G band damping due to scattering with a phonon[40]. The temperature-dependent resistivity of GD is well-fitted to the theoretical prediction considering the scattering with an acoustic and an optical phonon[18]. Scattering by geometric corrugations on a graphene surface is also an interesting subject. The corrugations can be formed by interaction with the substrate or by a thermal cycle during the annealing process[45]. From the theoretical calculations, electrons in the presence of a ripple are affected by the vector potential due to a strain field[37,46]. However, the role of corrugation in carrier transport remains not well understood.

3. Scanning Gate Microscope

The SGM in this study was operated under an ultrahigh vacuum (UHV) at 20 K with liquid helium, at 84K with liquid nitrogen, and at room temperature. The sample can be annealed with current through a device or annealing of the whole chamber can be accomplished at an elevated temperature. An ultra-sharp AFM tip can be positioned over the GD using custom-made vacuum motors. During the SGM operation, the transport current is measured through the electrodes, as shown in Fig. 1A. As we place the

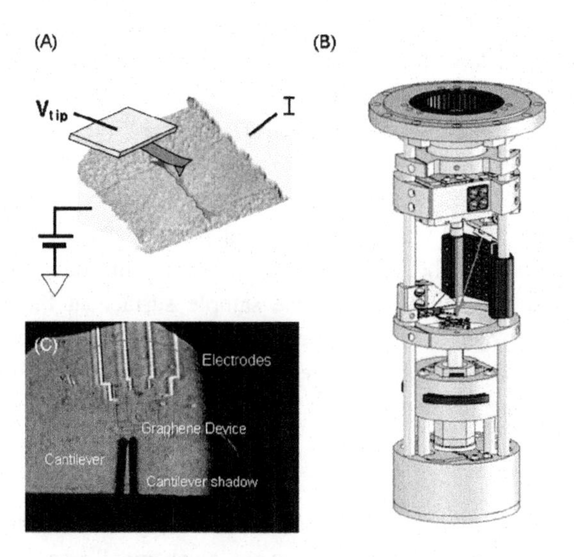

Fig. 1. (A) Schematic diagram of AFM operated at cryogenic temperature. This unit is kept inside a UHV tube that is inserted in a helium Dewar. A laser reflection from the back of a cantilever is detected by a quadrant photodiode. Piezoelectric motors are mounted to align the detection mechanism and the sample coarse movements. (B) CCD images of a GD with a long-range optical microscope installed inside UHV AFM chamber. (C) Operational principle of SGM. As an AFM cantilever scans a line, the controlling computer remembers the topography. After the scan a tip bias is applied to the tip and follows the topography, such that the tip follows the same distance away from the sample.

conducting AFM over the area of interest in a GD, the transport current varies as we move the device. A typical total conductance value without the tip gate is ~ 100 μS whereas the variation in the conductance with the tip gate is ~ 1 μS.

A custom-made UHV AFM operating at cryogenic temperatures was used to perform the SGM experiment, as shown in Fig. 1B[47]. The AFM head has a rigid three-column structure inside the UHV chamber, and the chamber is immersed in a liquid-helium Dewar and cooled by the helium exchange gas (at nearly one atmospheric pressure) to ensure optimal thermal and vibrational stability. The radiation through the main probe chamber is effectively cut off by internal baffles and the objective lens of the CCD optics is installed for a long-range optical microscope just above the AFM head. To achieve the maximum numerical aperture, we mounted an objective lens with a diameter of 25 mm and a focal length of 60 mm on top of the head. We drilled a small hole (4 mm diameters) in the lens along its optical axis to allow for the unhindered piezomotor-driven motion of the optical fibers. The intermediate image of the tip and the sample formed near the top of the main chamber is refocused onto the CCD by a telescopic microscope lens. Fig. 1C shows a long-range optical microscope image on the sample surface and the cantilever.

Vibration isolation is achieved by anchoring the chamber on a granite table. Three air table legs support the granite table, at 500 kg, and the cut off frequency is below ~3Hz. A helium Dewar is attached to three additional air dampers to isolate the mechanical vibration from the Dewar. The system is installed in a sound-proof room that is separate from the control and measurement circuits. The noise level from mechanical parts is much less than 100dB. A laser light bounced off the back of a commercial cantilever is detected by a quadrant photo diode to detect the z movement of the probe. The head has four inertial piezomotors: a one-dimensional (1D) Z motor[48] (10 mm span) for a coarse approach to the sample, and three planar XY motors (2 mm x 2 mm span each) for fine horizontal positioning of the sample and to adjust the laser lens assembly and the photodiode. A pan-type linear walker with six independent shear piezo-ceramic feet is used for coarse movement of the sample to the tip. As a white light source (the cleaved end of the multi-mode fiber) illuminates the sample surface at the angle θ, we can estimate the distance between the tip and the sample to be ten times the distance between the cantilever and its shadow laid on the sample, achieving a safe and fast coarse approach down to a tip-sample distance of ~ 5 μm.

4. Sample Preparation

Sub-headings The GDs used in this study were fabricated by fabrication processes similar to those reported earlier[4]. Graphene flakes were mechanically exfoliated from either highly oriented pyrolitic graphite (HOPG) or natural graphite. Graphite was tapped onto scotch tape and then peeled off several times until it appeared shiny. Subsequently, this graphene tape was transferred by finger onto the top of thermally grown 300-nm thick SiO_2 on a highly doped Si substrate that was used as back gate later. The thickness of the graphene flake was confirmed by micro-Raman spectroscopy. A single layer of graphene was confirmed via image contrast of an optical microscope. Source-drain metal (3 nm Cr / 30 nm Au) contacts were defined by conventional electron beam lithography and a lift-off process. Figure 2A shows an optical image of a fabricated GD with metallic contacts. These electrodes were used as markers to position the sample with the tip later, as shown in Fig. 1C. After the GD was mounted in a cryogenic AFM, a macroscopic transport measurement was performed to check the device characteristics, including the position of the Dirac point, electron and the hole mobility, as shown in Fig. 2B. Red and green lines represent data of a different sweep direction to note the hysteresis of the device. As the GDs were fabricated in an ambient environment and transferred to the AFM chamber, the macroscopic transport properties changed in the UHV chamber, as expected[49]. Figures 2C and D demonstrate the transport current as a function of the back gate bias voltage before the GD was inserted into a vacuum (2C) and in the vacuum (2D). Under ambient pressure, the Dirac point of the GD exceeded +20 V, indicating a hole-doped device by water molecule adsorption on graphene. After the GD was annealed at 380 K for 6 hours, the Dirac point moved to ~0 V, suggesting that the adsorbates were desorbed under UHV. The mobility of the sample was also increased when this process was repeated.

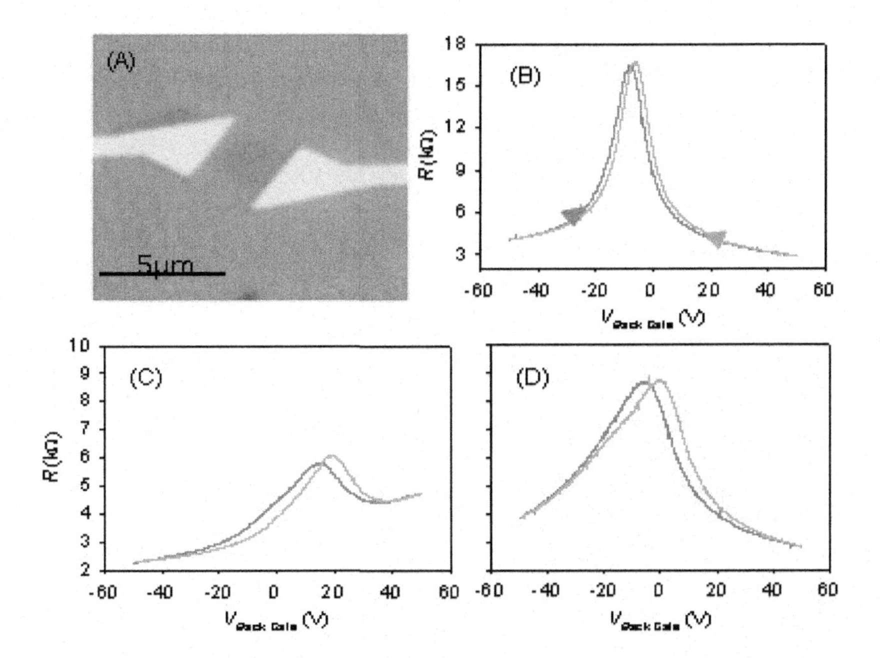

Fig. 2. (A) An optical microscope image of a fabricated GD. A faint dark object is a single- layer graphene and bright parts are contact electrodes. (B) A typical graphene resistance versus back gate bias voltage after loading into AFM and several annealing process. Red data was taken as increasing the bias and the green curve was obtained as ramping down. (C)-(D) Macroscopic transport measurement data before inserting into AFM chamber (C) and after annealing at 380K for 6 hours (D).

5. Results and Discussion

The topography of a graphene layer was measured using a scanning tunneling microscope (STM). In monolayer graphene, a honeycomb-shaped STM topography is observed. In double-layer graphene, every other carbon atom appears bright, resulting in a hexagonal topography due to the Bernal stacking of the double layers. Figure 3A shows an STM image of an epitaxially grown graphene layer on a Si-terminated SiC surface[50], showing the honeycomb-shaped atomic details. Imaging of the atomic details of a GD on SiO_2 was not possible with STM without an x-y motor or by the optical microscope in this chamber. In the SGM chamber, where x-y-z motors were equipped, it was possible to obtain a resolution of ~ 1 nm with a cantilever-based operation. Figure 3B shows a STS result on the graphene layer. It has a hump just below the Fermi level, as explained by phonon-mediated tunneling near the Dirac point[25].

After a GD was installed in a cryogenic AFM-SGM chamber, it was imaged in AFM mode. Figure 4A shows a contact mode AFM image of a GD, where a SiO_2-coated Si tip was used. The GD appears to have a lighter color in the middle. The corrugation on the SiO_2 surface is known to be ~1 nm with an average corrugation width of ~50-60 nm[51]. Similar corrugation was observed on the GD, and the characteristic features are very similar on the GD and a SiO_2 substrate. Figure 4B shows the AFM topography of a GD with a higher resolution. Again, the observed corrugation depth is very similar to that on a SiO_2 surface and the average separation among those corrugations is ~50-60 nm, as reported on a bare SiO_2 surface. These data suggest if there is a scattering mechanism due to corrugation, it can be modeled by this topography. Unlike earlier results[36], we could not find self-similarity in the length scale in our data.

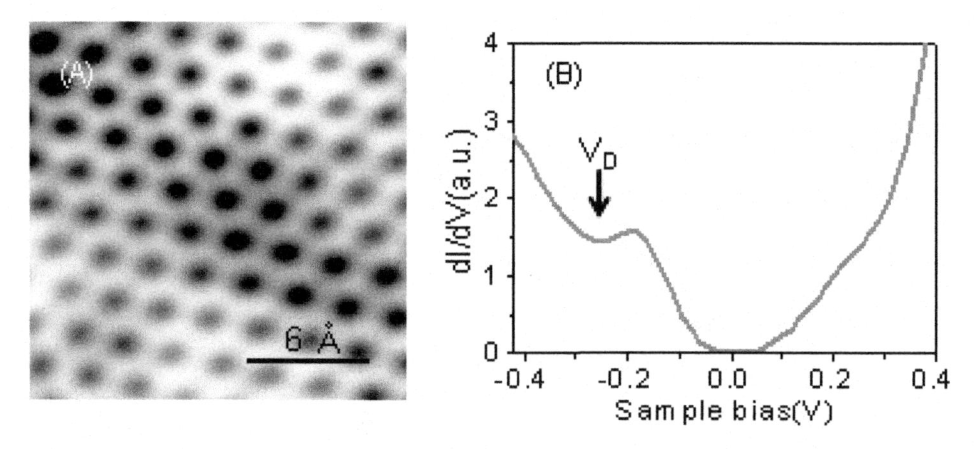

Fig. 3. (A) STM topography of a monolayer graphene on a Si-terminated SiC. (B) STS data on the surface of the monolayer graphene.

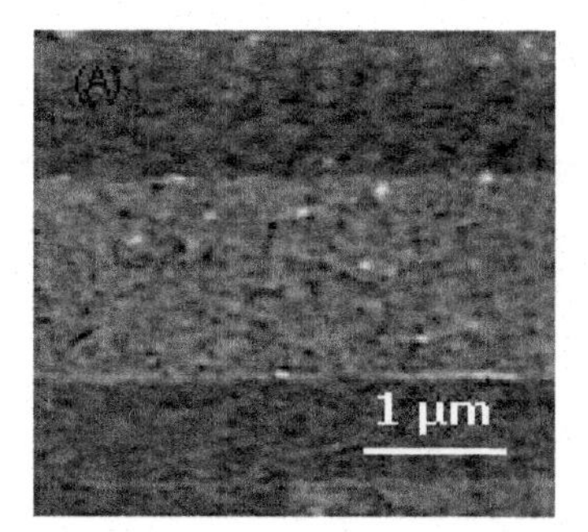

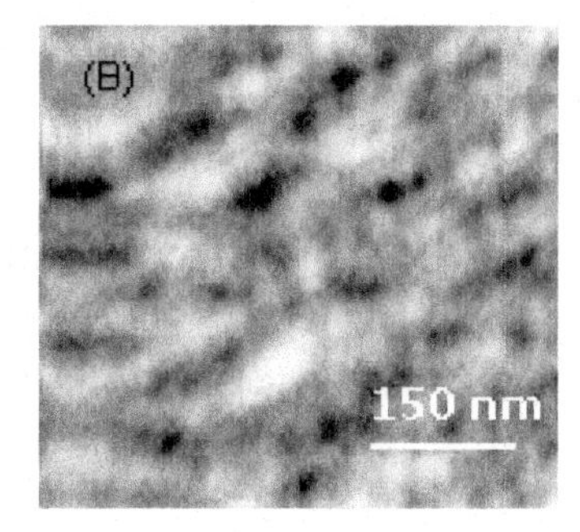

Fig. 4. (A) AFM topography of a GD measured in contact mode. (B) An AFM topography in non-contact mode. Maximum vertical corrugation is ~ 2 nm.

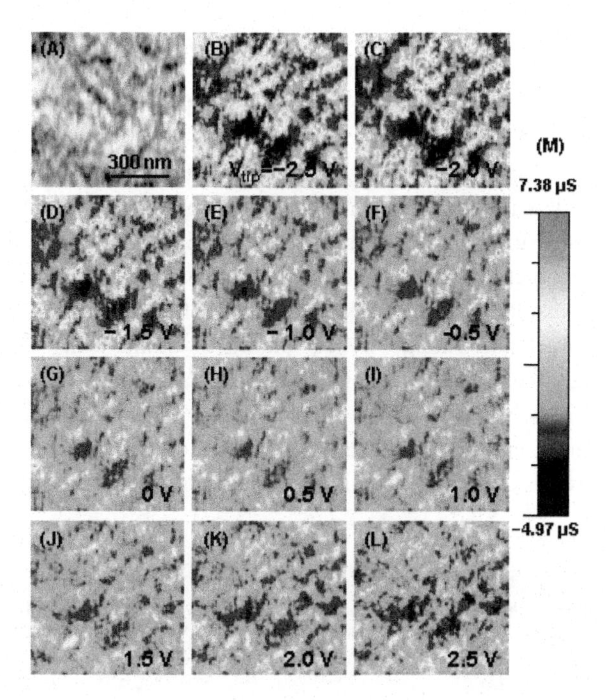

Fig. 5. (A) An AFM topography of 800 x 800 nm area on a GD. (B)-(L) SGM data on the same area at tip gating bias of −2.5 V, −2.0 V, −1.5 V, −1.0 V, −0.5 V, 0 V, 0.5 V, 1.0 V, 1.5 V, 2.0 V and 2.5 V respectively. (M) Color-coded scale bar of conductance change for SGM data.

After the corrugation was confirmed by AFM topography, SGM experiments were conducted. Figure 5A depicts a 800 x 800 nm^2 AFM topography of a GD and Figs. 5B-L show SGM micrographs from a tip bias voltage range of −2.5 to +2.5 in steps of 0.5 V. The Dirac point of this GD was −24 V, as set by the back gate bias. All of the SGM measurements were taken at Dirac point by back-gating. Corrugations on graphene surface are seen in topography image with lateral dimension of ~100 nm. The spatial variations of conductance change become more prominent as the absolute value of the tip gate bias increases in either negative or positive polarity. Negative (positive) polarity of electric field from the tip is related to the scattering dynamics of electron (hole) carriers. The small features in the SGM data are tens of nanometers, revealing good agreement with the AFM observation. There is strong correlation between the SGM signal and the topographic corrugations; the peaks appear at the bottom of the valley in topography. The range of high conductance region with hole carrier is extended more than that with electron carrier at the same electric field strength. This difference is due to the structure of charge puddle. The correlation between topographic data and SGM data implies that the charge puddles exist at the bottom of each ripple. The difference features in electron and hole carriers indicates that electrons are more locally confined at the bottom of the ripple induced by the interaction with substrate and hole carrier screens around confined electrons.

Unlike scattering by a geometrical corrugation, scattering caused by crystalline defects are slightly different. Figure 6A shows the topographic image with defect and Figs. 6B-F show the tip bias dependence of the SGM micrographs near defects with tip gating bias of −2.0 V (B), −4.0 V (C), 0 V (D), 2.0 V (E) and 4.0 V (F) respectively. The nature of this defect is not clear, but this area appears to be elastically deformed. The SGM signal indicates strong carrier scattering even at a high tip bias voltage[52] compared to scattering by geometrical corrugations in which the averaging effect by tip bias conceal the spatial resolution with tip gating voltage of higher than ±3 V. It is worth to note that the conductance change by tip gating is dramatically high compared to scattering by geometrical corrugations. Color-coded scale bar for SGM signal for defect is shown in Fig. 6G which is an order of magnitude higher than that for geometrical corrugation as shown in Fig. 5M. This result indicates that the defect can affect the transport properties in a dramatic way and has to be considered carefully.

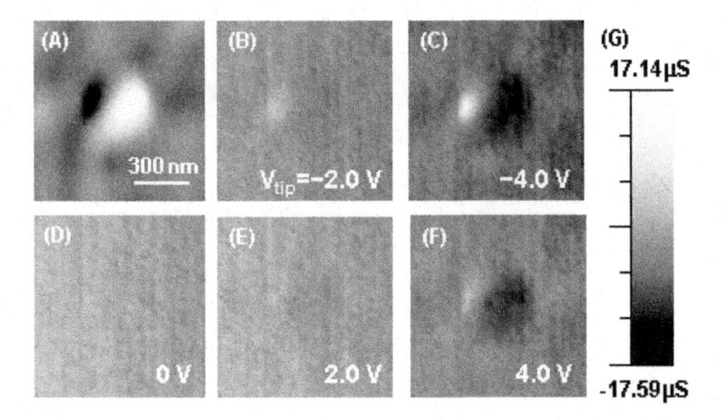

Fig. 6. (A) An AFM topography of geometrical defect. (B)-(F) Corresponding SGM signals with tip gating bias of −2 V, −4 V, 0 V, 2 V and 4 V respectively. (G) Color-coded scale bar of conductance change for SGM data.

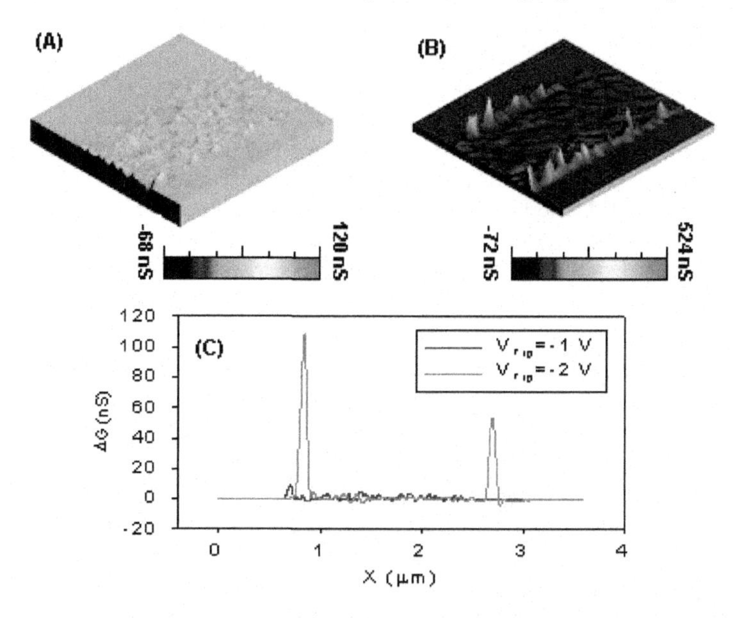

Fig. 7. SGM micrograph of a GD in which the Dirac point is -20 V of the back gate voltage with tip gate bias of −1 V (A) and −2 V (B). Strong conductance enhancement at the edge was observed with −2 V of tip bias which is more close to Dirac point.

When the SGM signal was measured at edges of a GD, a strong enhancement of the SGM signal was observed. Figures 7A and B show SGM results obtained by contact mode AFM with SiO_2-coated Si tip. The Dirac point of this sample was –20 V according to the back gate bias. As the tip gate bias was decreased to –2 V, Fig. 7B, i.e., as the tip gate approaches the Dirac point, the edge transport signal increases by as much as 10 times that at the middle of the GD. That is due to charge accumulation at the edge of a GD by electrostatic force[53]. The edge state conducting channel can be opened and contributed to electron transport like in a lightning rod. More experiment about edge effect with gate-controlled GD showed that the conductance enhancement at the edge was not observed with back gate of Dirac point (in low carrier density limit), however, the conductance enhancement at edge were observed again in electron or hole charging regime. Theoretical calculation with SGM simulation confirmed this edge effect by the edge channel opening[52]. Based on this result, it becomes necessary to re-interpret a considerable amount of existing macroscopic data.

In conclusion, this SGM study shows that geometric ripples, defects and edges directly influence the transport current. This observation is directly linked with the proposed scattering model based on macroscopic transport measurements, and the existing results must be re-examined.

This work was supported by National Research Foundation of Korea (3348-20090042).

References

1. K.S. Novoselov, A.K. Geim, S.V. Morozov, D. Jiang, Y. Zhang, S.V. Dubonos, et al., Electric Field Effect in Atomically Thin Carbon Films, Science. 306 (2004) 666-669.
2. K.S. Novoselov, A.K. Geim, S.V. Morozov, D. Jiang, M.I. Katsnelson, I.V. Grigorieva, et al., Two-dimensional gas of massless Dirac fermions in graphene, Nature. 438 (2005) 197-200.
3. Y. Zhang, Y. Tan, H.L. Stormer, P. Kim, Experimental observation of the quantum Hall effect and Berry's phase in graphene, Nature. 438 (2005) 201-204.
4. K.S. Novoselov, D. Jiang, F. Schedin, T.J. Booth, V.V. Khotkevich, S.V. Morozov, et al., Two-dimensional atomic crystals, Proceedings of the National Academy of Sciences of the United States of America. 102 (2005) 10451-10453.
5. A.H. Castro Neto, F. Guinea, N.M.R. Peres, Drawing conclusions from graphene, Physics World. 19 (2006) 33-37.
6. V.P. Gusynin, S.G. Sharapov, Unconventional Integer Quantum Hall Effect in Graphene, Phys. Rev. Lett. 95 (2005) 146801.
7. N.M.R. Peres, F. Guinea, A.H. Castro Neto, Electronic properties of disordered two-dimensional carbon, Phys. Rev. B. 73 (2006) 125411.
8. M.I. Katsnelson, K.S. Novoselov, A.K. Geim, Chiral tunnelling and the Klein paradox in graphene, Nat Phys. 2 (2006) 620-625.
9. M. Katsnelson, K. Novoselov, Graphene: New bridge between condensed matter physics and quantum electrodynamics, Solid State Communications. 143 (2007) 3-13.
10. A.F. Young, P. Kim, Quantum interference and Klein tunnelling in graphene heterojunctions, Nat Phys. 5 (2009) 222-226.
11. K. Bolotin, K. Sikes, Z. Jiang, M. Klima, G. Fudenberg, J. Hone, et al., Ultrahigh electron mobility in suspended graphene, Solid State Communications. 146 (2008) 351-355.
12. X. Du, I. Skachko, A. Barker, E.Y. Andrei, Approaching ballistic transport in suspended graphene, Nat Nano. 3 (2008) 491-495.
13. M. Orlita, C. Faugeras, P. Plochocka, P. Neugebauer, G. Martinez, D.K. Maude, et al., Approaching the Dirac Point in High-Mobility Multilayer Epitaxial Graphene, Phys. Rev. Lett. 101 (2008) 267601.
14. K.I. Bolotin, F. Ghahari, M.D. Shulman, H.L. Stormer, P. Kim, Observation of the fractional quantum Hall effect in graphene, Nature. 462 (2009) 196-199.
15. X. Du, I. Skachko, F. Duerr, A. Luican, E.Y. Andrei, Fractional quantum Hall effect and insulating phase of Dirac electrons in graphene, Nature. 462 (2009) 192-195.
16. D.L. Miller, K.D. Kubista, G.M. Rutter, M. Ruan, W.A. de Heer, P.N. First, et al., Observing the Quantization of Zero Mass Carriers in Graphene, Science. 324 (2009) 924-927.
17. Y. Tan, Y. Zhang, K. Bolotin, Y. Zhao, S. Adam, E.H. Hwang, et al., Measurement of Scattering Rate and Minimum Conductivity in Graphene, Phys. Rev. Lett. 99 (2007) 246803.
18. J. Chen, C. Jang, S. Xiao, M. Ishigami, M.S. Fuhrer, Intrinsic and extrinsic performance limits of graphene devices on SiO2, Nat Nano. 3 (2008) 206-209.
19. J. Chen, C. Jang, M. Ishigami, S. Xiao, W. Cullen, E. Williams, et al., Diffusive charge transport in graphene on SiO2, Solid State Communications. 149 (2009) 1080-1086.
20. F. Wang, Y. Zhang, C. Tian, C. Girit, A. Zettl, M. Crommie, et al., Gate-Variable Optical Transitions in Graphene, Science. 320 (2008) 206-209.
21. Z. Liu, K. Suenaga, P.J.F. Harris, S. Iijima, Open and Closed Edges of Graphene Layers, Phys. Rev. Lett. 102 (2009) 015501.
22. X. Jia, M. Hofmann, V. Meunier, B.G. Sumpter, J. Campos-Delgado, J.M. Romo-Herrera, et al., Controlled Formation of Sharp Zigzag and Armchair Edges in Graphitic Nanoribbons, Science. 323 (2009) 1701-1705.

23. C.O. Girit, J.C. Meyer, R. Erni, M.D. Rossell, C. Kisielowski, L. Yang, et al., Graphene at the Edge: Stability and Dynamics, Science. 323 (2009) 1705-1708.
24. G.M. Rutter, J.N. Crain, N.P. Guisinger, T. Li, P.N. First, J.A. Stroscio, Scattering and Interference in Epitaxial Graphene, Science. 317 (2007) 219-222.
25. Y. Zhang, V.W. Brar, F. Wang, C. Girit, Y. Yayon, M. Panlasigui, et al., Giant phonon-induced conductance in scanning tunnelling spectroscopy of gate-tunable graphene, Nat Phys. 4 (2008) 627-630.
26. E.J. Heller, K.E. Aidala, B.J. LeRoy, A.C. Bleszynski, A. Kalben, R.M. Westervelt, et al., Thermal Averages in a Quantum Point Contact with a Single Coherent Wave Packet, Nano Letters. 5 (2005) 1285-1292.
27. G. Metalidis, P. Bruno, Green's function technique for studying electron flow in two-dimensional mesoscopic samples, Phys. Rev. B. 72 (2005) 235304.
28. A. Freyn, I. Kleftogiannis, J. Pichard, Scanning Gate Microscopy of a Nanostructure Where Electrons Interact, Phys. Rev. Lett. 100 (2008) 226802.
29. M.G. Pala, B. Hackens, F. Martins, H. Sellier, V. Bayot, S. Huant, et al., Local density of states in mesoscopic samples from scanning gate microscopy, Phys. Rev. B. 77 (2008) 125310.
30. J.W.P. Hsu, N.G. Weimann, M.J. Manfra, K.W. West, D.V. Lang, F.F. Schrey, et al., Effect of dislocations on local transconductance in AlGaN/GaN heterostructures as imaged by scanning gate microscopy, Appl. Phys. Lett. 83 (2003) 4559.
31. B. Hackens, F. Martins, T. Ouisse, H. Sellier, S. Bollaert, X. Wallart, et al., Imaging and controlling electron transport inside a quantum ring, Nat Phys. 2 (2006) 826-830.
32. M.A. Topinka, B.J. LeRoy, R.M. Westervelt, S.E.J. Shaw, R. Fleischmann, E.J. Heller, et al., Coherent branched flow in a two-dimensional electron gas, Nature. 410 (2001) 183-186.
33. K.E. Aidala, R.E. Parrott, E.J. Heller, R.M. Westervelt, Imaging electrons in a magnetic field, Cond-Mat/0603035. (2006).
34. M.P. Jura, M.A. Topinka, L. Urban, A. Yazdani, H. Shtrikman, L.N. Pfeiffer, et al., Unexpected features of branched flow through high-mobility two-dimensional electron gases, Nat Phys. 3 (2007) 841-845.
35. N. Shon, T. Ando, Quantum Transport in Two-Dimensional Graphite System, J. Phys. Soc. Jpn. 67 (1998) 2421-2429.
36. M. Ishigami, J.H. Chen, W.G. Cullen, M.S. Fuhrer, E.D. Williams, Atomic Structure of Graphene on SiO2, Nano Letters. 7 (2007) 1643-1648.
37. M. Katsnelson, A. Geim, Electron scattering on microscopic corrugations in graphene, Philosophical Transactions of the Royal Society A: Mathematical, Physical and Engineering Sciences. 366 (2008) 195-204.
38. J. Chen, C. Jang, S. Adam, M.S. Fuhrer, E.D. Williams, M. Ishigami, Charged-impurity scattering in graphene, Nat Phys. 4 (2008) 377-381.
39. L.A. Ponomarenko, R. Yang, T.M. Mohiuddin, M.I. Katsnelson, K.S. Novoselov, S.V. Morozov, et al., Effect of a High- kappa Environment on Charge Carrier Mobility in Graphene, Phys. Rev. Lett. 102 (2009) 206603.
40. J. Yan, Y. Zhang, P. Kim, A. Pinczuk, Electric Field Effect Tuning of Electron-Phonon Coupling in Graphene, Phys. Rev. Lett. 98 (2007) 166802.
41. C. Jang, S. Adam, J. Chen, E.D. Williams, S. Das Sarma, M.S. Fuhrer, Tuning the Effective Fine Structure Constant in Graphene: Opposing Effects of Dielectric Screening on Short- and Long-Range Potential Scattering, Phys. Rev. Lett. 101 (2008) 146805.
42. J. Martin, N. Akerman, G. Ulbricht, T. Lohmann, J.H. Smet, K. von Klitzing, et al., Observation of electron-hole puddles in graphene using a scanning single-electron transistor, Nat Phys. 4 (2008) 144-148.

43. E. Rossi, S. Adam, S. Das Sarma, Effective medium theory for disordered two-dimensional graphene, Phys. Rev. B. 79 (2009) 245423.
44. Y. Zhang, V.W. Brar, C. Girit, A. Zettl, M.F. Crommie, Origin of spatial charge inhomogeneity in graphene, Nat Phys. 5 (2009) 722-726.
45. C. Chen, W. Bao, J. Theiss, C. Dames, C.N. Lau, S.B. Cronin, Raman Spectroscopy of Ripple Formation in Suspended Graphene, Nano Letters. 9 (2009) 4172-4176.
46. S.V. Morozov, K.S. Novoselov, M.I. Katsnelson, F. Schedin, L.A. Ponomarenko, D. Jiang, et al., Strong Suppression of Weak Localization in Graphene, Phys. Rev. Lett. 97 (2006) 016801.
47. J. Lee, J. Chae, C.K. Kim, H. Kim, S. Oh, Y. Kuk, Versatile low-temperature atomic force microscope with in situ piezomotor controls, charge-coupled device vision, and tip-gated transport measurement capability, Rev. Sci. Instrum. 76 (2005) 093701.
48. S.H. Pan, E.W. Hudson, J.C. Davis, [sup 3]He refrigerator based very low temperature scanning tunneling microscope, Rev. Sci. Instrum. 70 (1999) 1459.
49. T. Lohmann, K. von Klitzing, J.H. Smet, Four-Terminal Magneto-Transport in Graphene p-n Junctions Created by Spatially Selective Doping, Nano Letters. 9 (2009) 1973-1979.
50. H. Yang, A.J. Mayne, M. Boucherit, G. Comtet, G. Dujardin, Y. Kuk, Quantum Interference Channeling at Graphene Edges, Nano Letters. 10 (2010) 943-947.
51. X. Blasco, D. Hill, M. Porti, M. Nafria, X. Aymerich, Topographic characterization of AFM-grown SiO_{2} on Si, Nanotechnology. 12 (2001) 110.
52. J. Chae, S.Y. Jung, S.J. Woo, H.J. Yang, H. Baek, J. Hal, Y.J. Song, Y.-W. Son, N.B. Zhitenev, J.A. Stroscio, Y. Kuk, to be published (2010).
53. P.G. Silvestrov, K.B. Efetov, Charge accumulation at the boundaries of a graphene strip induced by a gate voltage: Electrostatic approach, Phys. Rev. B. 77 (2008) 155436.

WIDE BAND GAP TECHNOLOGY FOR HIGH POWER AND UV PHOTONICS

NOVEL APPROACHES TO MICROWAVE SWITCHING DEVICES USING NITRIDE TECHNOLOGY

G. SIMIN*, J. WANG, B. KHAN

Electrical Engineering, University of South Carolina, Columbia, SC, 29208, USA
**simin@cec.sc.edu*

J. YANG, A. SATTU, R. GASKA

Sensor Electronic Technology Inc., Columbia, SC, 29209, USA

M. SHUR

Electrical, Computer and System Eng., Rensselaer Polytechnic Institute, Troy, N.Y, USA

III-Nitride heterostructure field-effect transistors (HFETs) demonstrated a new paradigm in microwave switching and control applications due to unique combination of extremely low channel resistance (leading to low loss), very high RF power, low off-state capacitance, broad range of operating temperatures, chemical inertness and robustness. The paper reviews novel approaches and recent advances in III-Nitride technology for RF switching devices leading to higher operating frequencies and even lower insertion loss.

Keywords: Gallium nitride; Heterostructure Field-Effect transistors; Microwave Switches.

1. Introduction

Importance of RF switching and control devices in modern microwave systems continuously increases along with increase in their complexity and functionality. Software Defined Radio (SDR) systems, satellite communications, broad-band and multi-band mobile communications, and beam steering radars all require highly reliable, fast, low-loss RF switching components.

Currently, pin-diodes, GaAs MESFETs or HEMTs and RF MEMS are commonly used as RF switching and control components. All of these devices have significant performance limitations arising from the fundamental materials properties and/or device design such as significant forward currents and turn-off times in pin-diodes, high control voltages, self-actuation, low RF powers and reliability issues in RF MEMS, low RF powers in GaAs-based MESFETs and HEMTs.

AlInGaN based heterostructure field-effect transistors (HFETs) enable a new paradigm in the RF switch design. In these devices, the 2D electron gas having a record high density forms a metal-like conducting plate, which can be easily turned on and off by moderate gate control voltages. These features, in combination with extremely high breakdown voltages, chemical inertness and planar structure form a platform for excellent switching devices. The insulated gate device design (MOSHFET) further develops these features into reliable and robust superior RF switch with ultra-low control powers, high linearity and high RF powers. RF switches using III-Nitride MOSHFETs can operate with 10 - 100 times higher power of RF signals as compared to the available RF MEMS and GaAs HEMTs. They have two-three orders of magnitude lower power consumption and switching times than those of pin-switches. They also have high reliability and thermal stability with operating temperatures ranging from cryogenic temperatures up to 300°C or even higher. These features make the III-Nitride technology to be the best candidate to replace most of the existing RF switching components. The achieved results confirm great potential of III-N MOSHFETs as ultimate RF switching devices and MMICs [1-6]

2. Key III-N Switch Performance Parameters and Limitations

In a typical series transistor switch configuration, the source and drain electrodes are connected to the RF line input and output correspondingly. The gate electrode is connected to a control voltage supply through a blocking resistor. Fig. 1 shows the equivalent circuit of the series FET RF switch, including the variable, gate-voltage controlled channel resistance, parasitic device capacitances and inductors associated with the bonding wires.

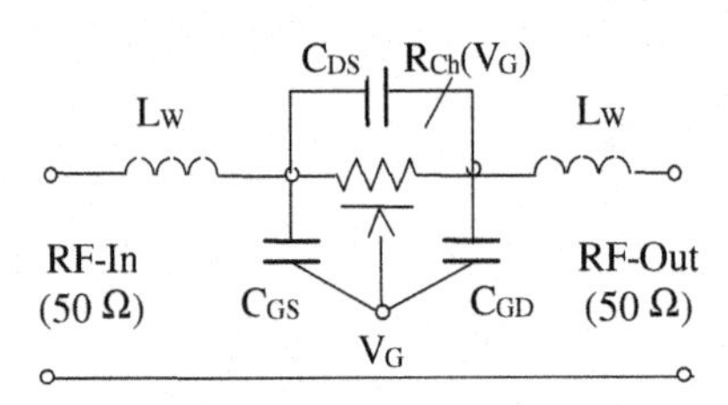

Fig. 1. Equivalent circuit of a series FET switch including device parasitic capacitances and mounting wire inductances (after[7]).

In the "ON" state, the HFET or MOSHFET switch gate bias is zero or positive, the channel resistance R_{ON} is low that ensures low-loss input-output transmission. The drain-source capacitance is shunted by the R_{ON}, therefore the transmission is almost frequency independent. The value of the R_{ON} can be estimated as $R_{ON} = 2R_C + R_{GS}+R_{GD} + R_G$, where R_C is the contact resistance, $R_{GS} = R_{SH}\times L_{GS}/W$ and $R_{GD} = R_{SH}\times L_{GD}/W$, $R_{SH} = 1/(qN_S\mu_n)$ is the layer sheet resistance, W is the device width, N_S is the channel sheet electron density, μ_n is the channel electron mobility, L_{GS}, L_{GD} are the source-gate and gate-drain spacing, and R_G is the voltage dependent resistance of the channel under the gate. At zero or positive gate bias $R_G \approx R_{SH}\times L_G/W$, where L_G is the gate length. For a typical MOSHFET, $R_{SH} \approx 300$ Ω/sq, $L_G = 1$ μm, $L_{GS} \approx L_{GD} \approx 1.5$ μm, the contact resistance, $R_C \approx 0.5$ Ω×mm, $R_{ON} \approx 2.2$ Ω×mm. The insertion loss of a series resistor R_{ON} connected into a transmission line with the characteristic impedance Z_0 ($Z_0 = 50$ Ω typically), assuming $R_{ON} << Z_0$, can be found as $IL(dB) = -20\,Log\frac{1}{1+R_{ON}/2Z_0} \approx 0.087\,R_{ON}$. For example, for the MOSHFET with the electrode width W = 1 mm, the insertion loss is as low as 0.2 dB and decreases inversely proportional to the width W.

At high frequencies, the isolation of a series switch is limited by device capacitance in the off-state. Referring to Fig. 1 and noting that $C_{GS} \approx C_{GD}$, the off-capacitance can be found as $C_{OFF} \approx C_{DS}+C_{GS}/2$. For a typical MOSHFET geometry described above, according to experimental and simulated data, $C_{DS} \approx 0.05$ pF/mm, $C_{GS} \approx 0.18$ pF/mm and $C_{OFF} \approx 0.14$ pF/mm. The isolation can be found as $IS(dB) = 20\,Log\dfrac{1}{\left|1+1/j\omega C_{OFF}2Z_0\right|}$.

For W = 1 mm and f = 2GHz, IS $\approx$ 18 dB.

The isolation can be significantly improved using series-shunt configuration, see Fig. 2.

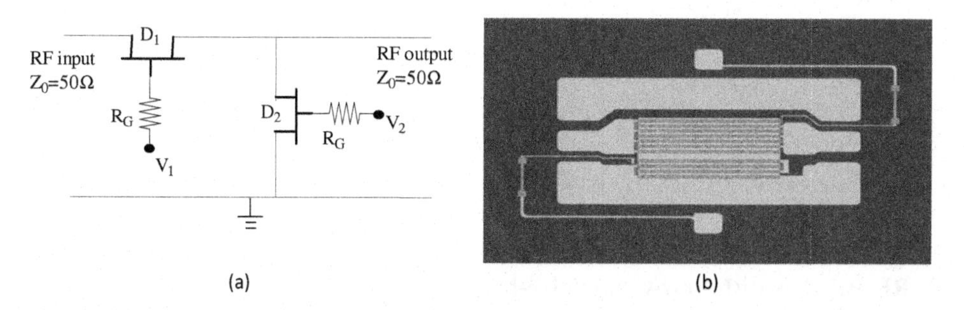

Fig. 2. Equivalent circuit with series (D1) and shunting (D2) HFETs (b) CCD image of low-loss SPST RF switch MMIC (after [4]).

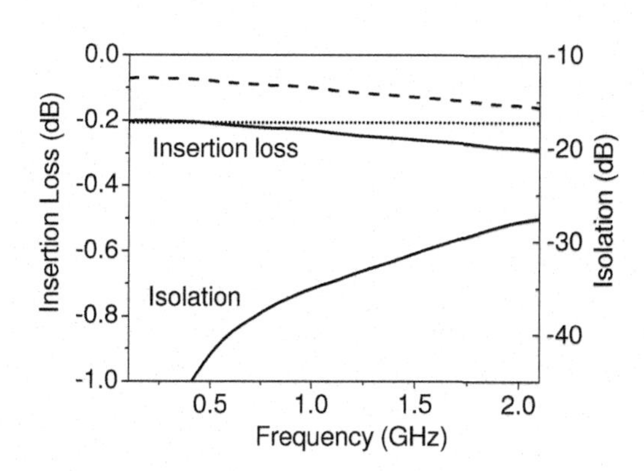

Fig. 3. Frequency dependencies of insertion loss and isolation of the series-shunt single-pole-single throw (SPST) switch. Solid lines – experimental data. Dotted line shows the simulated insertion loss in the absence of shunt HFET capacitance. Dashed line shows the simulated insertion loss in the absence of loss in the metal electrodes (after [4]).

Using series-shunt switch layout, one can simultaneously achieve low insertion loss and high isolation as illustrated in Fig. 3. Excessive loss of around 0.13 dB, as compared to the simulated data, comes from the resistance of metal electrodes; this can be eliminated by increasing the metal thickness, e.g. using electroplated metal electrodes. The frequency dependence of the insertion loss comes from the off-capacitance of the shunt device. As seen, even in series-shunt configuration, the device off-capacitance C_{OFF} remains the key factor limiting the RF switch bandwidth. Because C_{OFF} is proportional to the device width W, reducing C_{OFF} using narrower devices will lead to higher R_{ON} values and hence higher insertion loss and lower isolation.

Major efforts in the RF switch technology development are focused on reduction of the $R_{ON}C_{OFF}$ product, which is the important figure of merit for most of RF switching devices. Another direction is to overcome the limitations imposed by C_{OFF} using innovative circuit layouts.

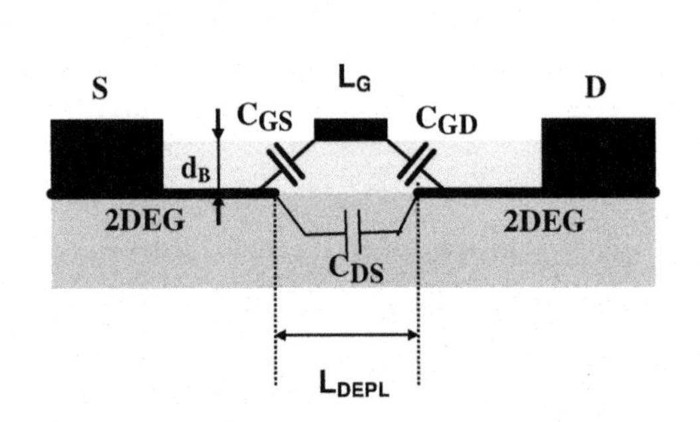

Fig. 4. Equivalent circuit of HFET in the OFF state.

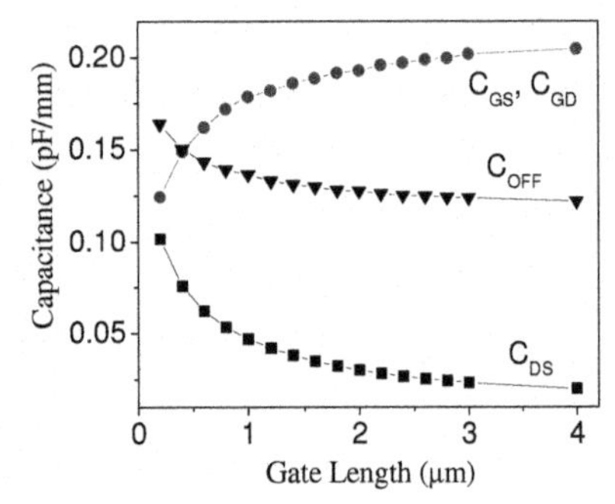

Fig. 5. HFET OFF-capacitance components – gate length dependencies.

3. RF Switches with Low-Conductive-Layer Coating

When HFET (or MOSHFET) is in the OFF-state, the 2D channel is fully depleted under the gate but still exists in the source-gate and gate-drain openings (Fig. 5). In regular HFET structures, the spacing between the 2DEG layer edges is therefore close to the gate length L_G.

The components of the OFF-state capacitances of conventional AlGaN/GaN HFETs have been simulated using the ANSOFT Maxwell 2D simulator. The 2D channel was modeled by a thin metal plane. The depletion region extensions into the gate-source and gate-drain openings were taken as d_B each. The source-gate spacing was L_{SD} = 5μm, the source and drain electrode length was 24 μm each. The C_{GD}, C_{GS} and C_{DS} partial capacitance and the total C_{OFF} capacitance dependencies on the gate length L_G are shown in the figure Fig. 5. As seen, changing the gate length leads to decreasing in the C_{DS} capacitance with simultaneous increase in the C_{DG} and C_{GS}. As a result, the total capacitance C_{OFF}, that is calculated using an equivalent circuit of Fig. 4, does show any significant dependence on the gate length.

We now consider the HFET with an additional low-conducting layer (LCL) under the gate. When the bias is applied at the gate electrode, the potential along LCL in the steady state equals to that of the gate. Therefore, in the off-state the 2DEG is removed in the entire region under the LCL and gate electrode. If the LCL layer conductivity is low, the shunting effect of it on the RF signal leak from the source to the drain can be minimal. The LCL can therefore reduce the off-capacitance without degrading the off-resistance. The quantitative estimate of the effect of LCL on the off-capacitance was obtained by 2D simulations similar to those used to obtain the Fig. 5. We varied the length of the 2DEG depletion region from that of the gate length L_G up to the entire source- drain spacing,

L_{SD}. Fig. 7 shows the simulation results for two device layouts with different electrode length and electrode spacing. The black line corresponds to the typical current technology switches and the red line corresponds to the layout with smaller ohmic contact length (note that the term "contact length" here refers to the electrode size in the direction from source toward drain. One can see that increasing the length of the depletion region between the source and drain sides of the 2DEG layers results in significant reduction in the OFF-capacitance. The reduction is around 1.9 times for 24 μm-long source and drain and around 2 times for 4 μm-long electrodes. In the latter case the C_{OFF} reduction is more pronounced because of lower source - drain electrode capacitance.

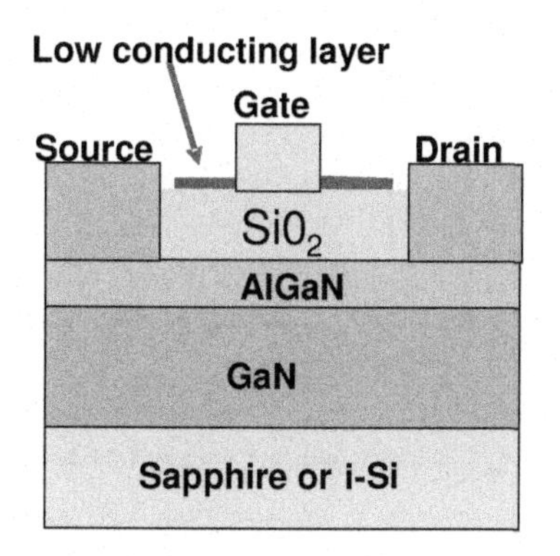

Fig. 6. MOSHFET with LCL under the gate.

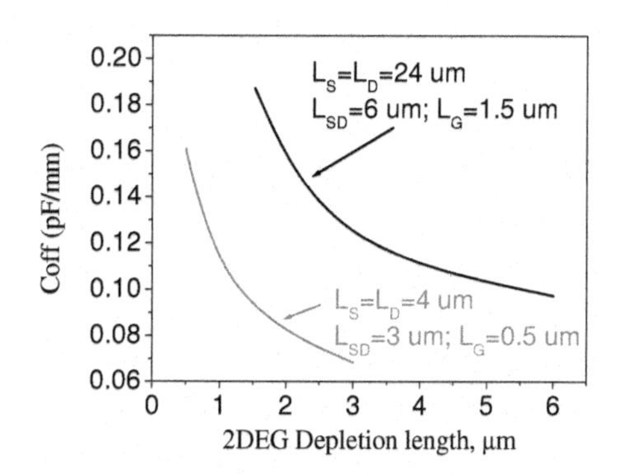

Fig. 7. Off-state capacitance of the HFET with the 2Delectron gas removed within the distance L_{DEPL}.

In order to formulate the requirements to the LCL resistance, we consider the LCL covering the entire source – drain spacing. This layer creates a parasitic conducting path between the source and drain which limits the achievable isolation. Let us require that the isolation provided by this additional conductance is at least 30 dB.

The resistance of the surface LCL between the source and the gate can be found from the following expression: $IS(dB) = 20\,Log\,\frac{1}{|1 + R_{SURF}/2Z_0|}$.

From this, the surface resistance of the LCL should not be lower than $R_{SURF} \approx 3.1$ kΩ. To convert this value into a more practical characteristic of the gate material – the surface sheet resistance, assume the source-drain spacing of L_{SD} = 5 μm and the total width of W = 3 mm (this width is typical for the switch with insertion loss less than 0.1 dB). $R_{SH} = R_{SURF} \times W/L_{SD} \approx 2$ MΩ/sq. In case the LCL is not in direct contact with the source and drain electrodes, the required resistance could be less than the above without isolation degradation.

The above criteria have been used in choosing the surface cladding material and developing the technology to depositing the surface conducting layer. MOSHFET switches with LCL cladding have been fabricated using low-temperature MOCVD deposited InGaN layer. The LCL cladding was deposited on top of the oxide layer before the gate metallization. The comparison between fabricated series-shunt single-pole double-throw switches with and without InGaN LCL cladding is shown in Fig. 8. As seen, at low frequencies, both switch type show identical insertion loss and isolation. However at high frequencies, above ~ 3 GHz, the switch with LCL shows lower insertion loss and higher isolation. At 6 GHz, the insertion loss of LCL-coated switch is 0.6 dB lower and the isolation is ~ 5 dB higher. This result confirms significant reduction of the off-capacitance in MOSHFET with low-conducting layer under the gate and validates the proposed approach.

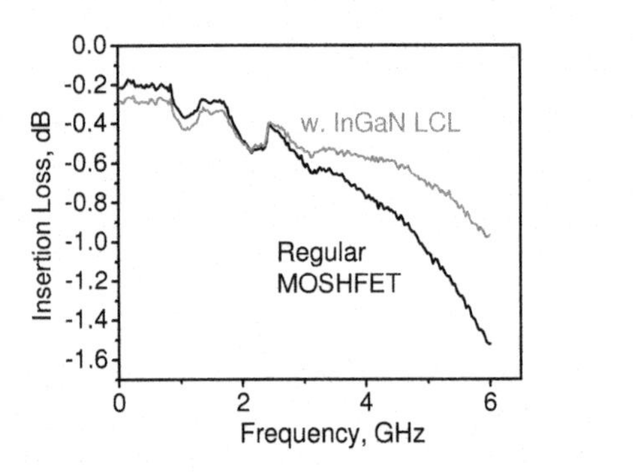

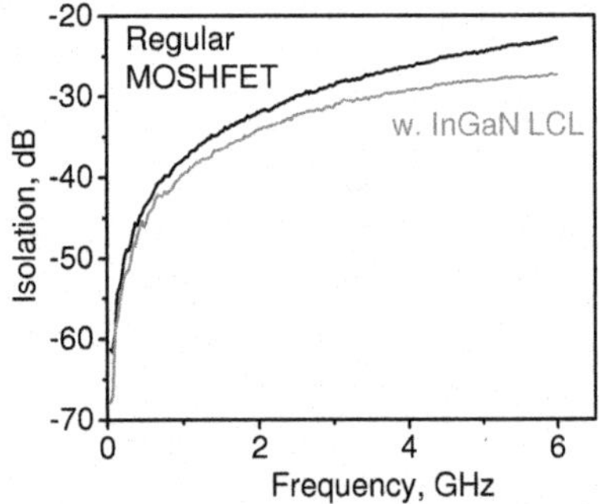

Fig. 8. Insertion loss and isolation of series-shunt MOSHFET SPDT switches with and without InGaN LCL.

4. Gateless Traveling Wave Switch [8]

One of the most efficient approaches to compensating the capacitive component of the impedance is realized in traveling wave switch (TWS) configuration where the device capacitance becomes a part of a transmission line and thus does not limit the bandwidth. [9,10] Typical TWS consists of a number of shunt transistors which resistance and capacitance are controlled by the gate voltage. To achieve a broadband operation, a large number of elementary transistor cells is required, each cell being a multi-finger transistor. In traditional microwave switch technologies, practical number of elementary cells is limited due to tight source-drain spacing and gate alignment issues. We present a novel design for low-loss, high-power TWS using elementary cells formed by voltage controlled AlGaN/GaN capacitors. In these devices, the control electrode is located outside the RF channel. As shown in Fig. 9, the depletion of the 2D electron gas channel under the input and output RF electrodes is achieved by applying a positive bias between the control electrode and two capacitively-coupled contacts (C^3) forming RF input and output electrodes. This novel design offers significant advantages over conventional field-effect transistor switches. The absence of the gate radically simplifies switch fabrication, especially in multi-element layout required for traveling wave devices. The gateless design also allows for smaller electrode spacing and thus leads to a lower ON-resistance.

Fig. 10 shows the C^3-TWS layout with unit cell details shown in the insert. The transmission line is formed by high-impedance coplanar waveguide lines and shunt-connected C^3 electrodes. The TWS was fabricated using AlGaN/GaN heterostructures on sapphire substrates. The C^3 cells were formed by mesa etching using RIE. To achieve low leakage, the electrodes were isolated from the AlGaN surface by a 15 nm thick SiO_2 layer deposited by PECVD. An additional layer of SiO_2 was used for dielectric bridges to isolate the control voltage bus from the RF lines. Two versions of the C^3-TWSs = 890 μm long (10 unit cells) and 1780 μm long (20 unit cells) - have been fabricated and tested. The experimental data are shown in Fig. 11. As seen, the TWS has an excellent combination of low insertion loss (0.6 – 1.2 dB) and high isolation (40 – 60 dB) at 20 GHz. The 890 μm long TWS was also fabricated and tested with the integrated quarter wave length transformer connected between the switch and the RF input. Large signal performance has been tested using the input powers up to 40 dBm. Less than 1 dB compression has been observed in this power range.

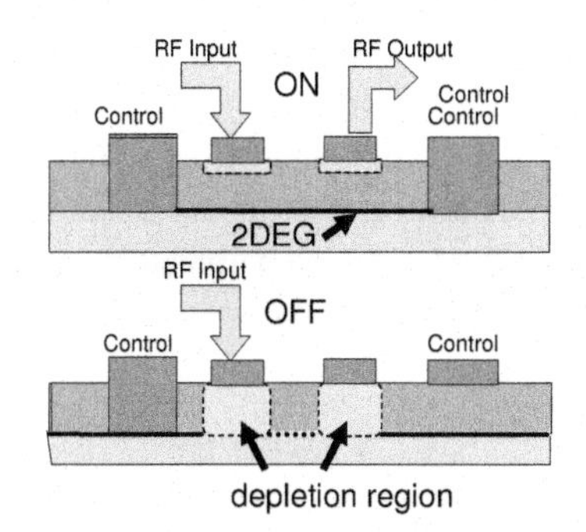

Fig. 9. Gateless microwave switch with capacitively-coupled contacts - operation concept.

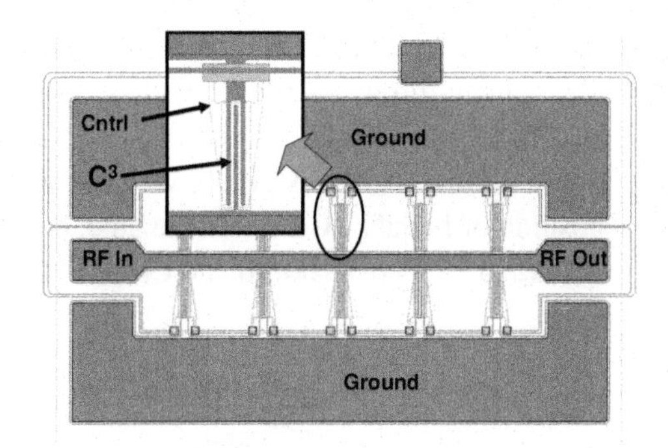

Fig. 10. Schematic layout of the TWS with capacitively-coupled contacts over the SiO_2/AlGaN/GaN heterostructure.

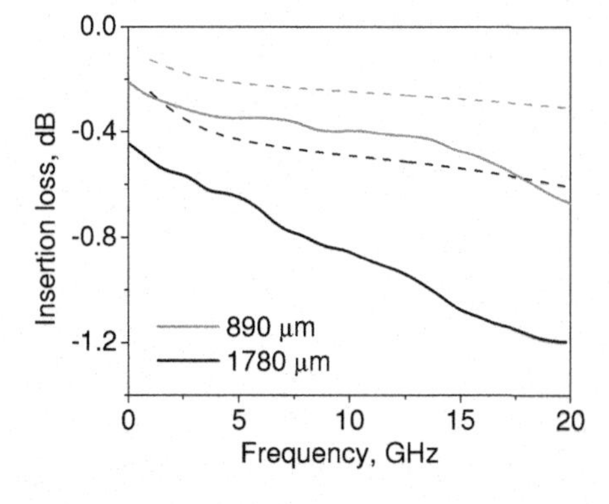

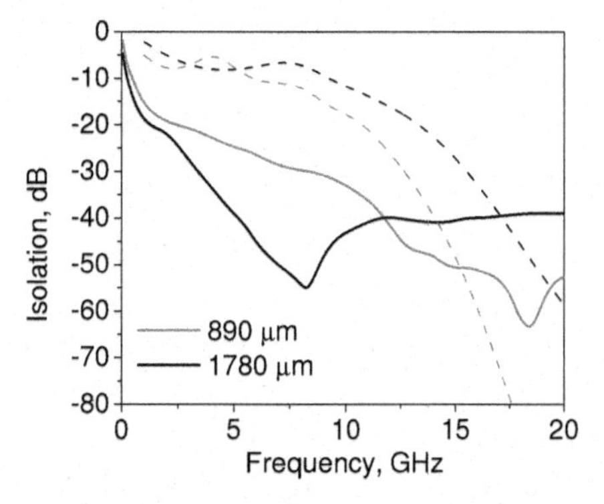

Fig. 11. Insertion loss and isolation of two C3-TWSs of different length. The 890 μm and 1780 μm long switches contain 10 and 20 unit cells correspondingly. The On state is achieved by applying +10V bias to the control electrode. The Off-state is measured at zero control voltage. The dashed lines show the simulated data.

5. AlInN/GaN Heterostructures for RF Switches

ON-resistance is a critical parameter of microwave switches strongly affecting the achievable insertion loss and isolation. As compared to AlGaN/GaN materials, lower R_{ON} values can be obtained in AlInN/GaN heterostructures with high Al composition. We used Migration Enhanced MOCVD (MEMOCVD®) deposition technology [11] for growing high quality AlInN/AlGaN/GaN heterostructures with In fraction varied from 12% to 22% (In molar fraction of 17-18% corresponds to the lattice-matched condition to GaN buffer.) The typical sheet carrier density was $2.45\times10^{13}cm^{-2}$ and minimal sheet resistance was close to 200 Ω/sq. In addition to the high sheet carrier densities, the contact resistance as low as 0.07 Ωxmm was obtained, which we attribute to thin barrier layer and high In-composition in it.

Using the material data for AlGaN/GaN and AlInN/GaN heterostructures, we performed an analysis of the best achievable performance for various switch configurations using these two material types. Fig. 12 shows comparative results for single-pole-single-throw (SPST) switches. For AlGaN/GaN material we assumed the sheet resistance R_{SH} = 300 Ω/sq, contact resistance R_C = 0.5 Ωxmm and off capacitance C_{OFF} = 0.14 pF/mm as discussed above. For the AlInN/GaN structures, we also assumed the deposition of InGaN LCL under the gate; correspondingly the design parameters were R_{SH} = 200 Ω/sq., R_C - 0.1 Ωxmm and C_{OFF} = 0.07 pF/mm.

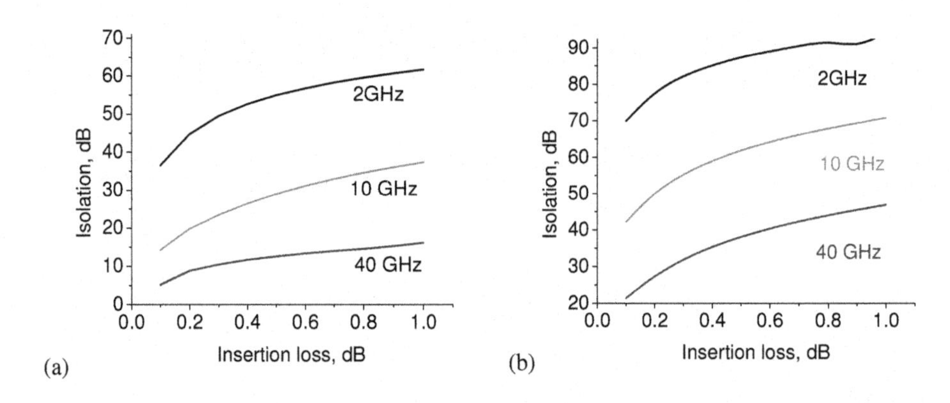

Fig. 12. Best achievable performance for SPST switch using conventional AlGaN/GaN heterostructures (a) and AlInN/GaN heterostructures in combination with InGaN LCL cladding (b).

The results presented in Fig. 12, demonstrate the achievable performance using the combination of innovative technologies discussed in this paper. As seen, introducing novel heterostructures and measures to reducing the off-state capacitance allow achieving impressive performance up to 40 GHz. As an example the combination of insertion loss of 0.5 dB and isolation of 35 dB is achievable. Noting that in addition to low loss and high isolation, the III-Nitride switches also demonstrated outstanding high-power, high-temperature characteristics, we conclude that these innovations make III-Nitride switches the best performing devices among nearly all other types of solid-state switches.

Acknowledgments

This work was supported by the NSF grant monitored by Dr. P. Fulay (USC), ONR grant monitored by Dr. P. Maki (RPI, USC) and DARPA contract monitored by Dr. S. Raman (SET Inc., USC, RPI).

References

1. H. Ishida, Y. Hirose, T. Murata, Y. Ikeda, T. Matsuno, K. Inoue, Y. Uemoto, T. Tanaka, T. Egawa, and D. Ueda, "A high-power RF switch IC using AlGaN/GaN HFETs with single-stage configuration", IEEE Trans. Electron Devices, vol. 52, pp. 1893-1899, 2005
2. G. Simin, X. Hu, Z. Yang, J. Yang, M. Shur, and R. Gaska, Low-Loss High-Power AlInGaN RF Switches, ISDRS 2007, International Semiconductor Device Research Symposium Dec. 12-14, 2007, University of Maryland, College Park, MD
3. G. Simin, High-power III-Nitride microwave devices with capacitively-coupled electrodes, WOCSEMMAD '07, The Workshop on Compound Semiconductor Materials and Devices, February 18-21, 2007, The Mulberry Inn, Savannah, Georgia
4. Z. Yang, J. Wang, X. Hu, J. Yang, G. Simin, M. Shur, and R. Gaska, Current Crowding in High Performance Low-Loss HFET RF switches, IEEE El. Dev. Lett. V. 29, 15-17 (2008)
5. G. Simin, M. Asif Khan, M. S. Shur, and R. Gaska, High-Power Switching Using III-Nitride Metal-Oxide-Semiconductor Heterostructures. Book Chapter in: Selected Topics in Electronics and Systems - Vol. 41. FRONTIERS IN ELECTRONICS Edited by H Iwai, Y Nishi, M S Shur, and H Wong. ISBN 981-256-884-0
6. Z. Yang, A. Koudymov, V. Adivarahan, J. Yang, G. Simin, and M. A. Khan, High-Power Operation of III-N MOSHFET RF Switches IEEE Microwave and Wireless Components Lett., VOL. 15, NO. 12, DECEMBER 2005, 850-852
7. G. Simin, M. Asif Khan, M. S. Shur, and R. Gaska, High-Power Switching Using III-Nitride Metal-Oxide-Semiconductor Heterostructures. Int. Journ. of High Speed Electronics and Systems v.16, N2, p.455-468 (2006)
8. G. Simin, M. Shur, and R. Gaska, New III-Nitride Technology For Microwave Switching, paper WoReSim0134, 34th WOCSDICE, May 17-19, 2010, Darmstadt/Seeheim, Germany
9. H. Mizutani and Y. Takayama, DC–110-GHz MMIC Traveling-Wave Switch, IEEE Trans MTT, V. 48, 840-845 (2000)
10. S. Chang, W. Chen, J. Chen, H. Kuo, and H. Hsu, New Millimeter-Wave MMIC Switch Design Using the Image-Filter Synthesis Method, IEEE MWC Lett, V. 14, 103-105 (2004)
11. Q. Fareed, R. Gaska, and M. S. Shur, Methods of Growing Nitride-Based Film Using Varying Pulses, US Patent 7,192,849, March 20 (2007)

AUTHOR INDEX

Bae, Y-H. 81
Baek, H. 205
Balandin, A. A. 161
Bawedin, M. 81
Bayraktaroglu, B. 171
Belenky, G. 43
Bengtsson, S. 195
Bersuker, G. 27, 65, 95
Birner, S. 143

Campbell, E. E. B. 195
Casse, M. 81
Chae, J. 205
Chang, S-J. 81
Chen, J. 43
Chien, L. H. 143
Cristoloveanu, S. 81

Dai, J. 3
Denorme, S. 81
Dibiccari, M. 13
Dobrinsky, A. 153
Donais, C. 27

Ek-Weis, J. 195
Engström, K. 195
Enoksson, P. 195
Eriksson, A. 195

Faynot, O. 81
Fenouillet-Beranger, C. 81

Gao, Q. 131
Gaska, R. 219
Ghavanini, F. A. 195
Ghibaudo, G. 81
Goel, N. 95

Ha, J. 205
Haider, M. R. 115
Heeg, T. 105
Ho, I. C. 3
Hobbs, C. 27
Holleman, J. 115
Hosoda, T. 43
Hou, W. 13

Ionica, I. 81
Islam, S. K. 115

Jackson, H. E. 131
Jagadish, C. 131
Jammy, R. 27, 65
Joyce, H. J. 131
Jung, S. Y. 205

Kambhampati, R. 95, 105
Kang, J. H. 131
Khan, B. 219
Kim, Y. 131
Kipshidze, G. 43
Kirsch, P. 65
Koveshnikov, S. 95, 105
Kuk, Y. 205

Larcher, L. 65
Leedy, K. 171
Li, Y. 13
Liang, R. 43
Liu, G. 161
Liu, J. 3
Lu, X. 3
Lundgren, P. 195

Mitin, V. 143

Morassi, L. 65
Mostafa, S. 115
Mynbaev, D. K. 51

Nagaiah, P. 95
Neidhard, R. 171

Ohata, A. 81
Oktyabrsky, S. 95, 105
Olofsson, N. 195

Padovani, A. 65
Paiman, S. 131
Park, H. 65
Park, K-H. 81
Perreau, P. 81
Pham-Nguyen, L. 81

Qin, L. 183

Rumyantsev, S. 27, 105, 161

Sadrzadeh, A. 153
Sattu, A. 219
Sawyer, S. 183
Schlom, D. G. 105
Sergeev, A. 143
Shing, C. 183
Shterengas, L. 43
Shur, M. 27, 105, 161, 219
Simin, G. 219
Skotnicki, T. 81
Smith, C. 27
Smith, L. M. 131
Son, Y.-W. 205
Song, Y. J. 205
Stark, C. 13
Stillman, W. 27, 105, 161
Stroscio, J. A. 205
Sukharenko, V. 51

Tan, H. H. 131
Taylor, W. 27, 65
Tokranov, V. 95, 105
Tsvid, G. 43

Vagidov, N. 143
van den Daele, W. 81
Veksler, D. 27, 65, 95

Wang, J. 219
Westerfeld, D. 43
Wetzel, C. 13
Woo, S. J. 205

Xia, Y. 13
Xu, J. 153

Yakimov, M. 95
Yakobson, B. I. 153
Yang, H. J. 205
Yang, J. 219
Yarrison-Rice, J. M. 131
You, S. 13
Young, C. D. 65

Zhang, X. C. 3
Zhao, L. 13
Zhao, W. 13
Zhitenev, N. B. 205
Zhu, M. 13
Zou, J. 131